Department of Bacteriology,
Westminster Medical School.

CLINICAL
CHEMICAL PATHOLOGY

CLINICAL CHEMICAL PATHOLOGY

Department of Bacteriology,
Westminster Medical School

C. H. GRAY
D.Sc., M.D., F.R.C.P., M.R.C.S., F.R.I.C., F.C.Path.
Professor of Chemical Pathology, University of London
Chemical Pathologist, King's College Hospital, London

Fifth edition

EDWARD ARNOLD (Publishers) LTD., London

© C. H. GRAY, 1968

First published 1953
Reprinted   1955
Second edition 1959
Reprinted   1961
Third edition   1963
Fourth edition 1965
Reprinted   1967
Fifth edition   1968

SBN: 7131 4134 4

PRINTED IN GREAT BRITAIN BY
RICHARD CLAY (THE CHAUCER PRESS), LTD.,
BUNGAY, SUFFOLK

PREFACE TO THE FIFTH EDITION

THE chapter on the biochemical tests in endocrine diseases has been re-written and considerable recent work has been introduced into the chapters on liver function and plasma proteins. All except five of the remaining chapters have received minor modifications.

The ever expanding field of chemical pathology has suggested to the author that the time has come when the task of keeping this book up-to-date should be shared by one or more co-authors, but I have been persuaded that this might be undesirable by numerous kind friends. I gratefully acknowledge my indebtedness to Professor Nicholas Martin, Dr. Roger Williams and Dr. Dudley Tee for valuable comments and criticisms, to Professor V. H. T. James, Dr. Peter Dixon and Dr. Mollie Booth for much discussion and help with the sections on adrenocortical function, to Dr. Anthony Ames for the preparation of Table 11, and to Dr. R. Burton for preparing the addendum to Chapter II. Permission is acknowledged from Drs. Kerr and Hall and Messrs. J. and A. Churchill for their permission to make use of Table 10 which is a modified form of a table in *Progress in Clinical Medicine*, 1966, from Dr. Davenport and The University of Chicago Press for Figure 3(a), which is a modification of a figure which appears in *The ABC of Acid-Base Chemistry* and from Drs. Siggard and Astrup to use Figure 3(b).

I am again grateful to Miss Patricia Thornton for the patience with which she has prepared numerous versions of the typescript and index and others of the staff of the Chemical Pathology Department for help in reading the proofs.

C. H. G.

PREFACE TO THE FIRST EDITION

THIS short text is based on a series of lectures in chemical pathology given to the medical students at King's College Hospital Medical School. It does not pretend to be comprehensive, but presents those features of the subject which seem to the author to be of particular value in assisting the medical student to appreciate some of the chemical aspects of disease.

Nevertheless it is hoped that this account will reach a wider circle of readers than clinical medical students, and that it may be of value to housemen and registrars in providing a readily available, yet concise, book of reference. Perhaps, indeed, consultants and general practitioners may find some useful information in these pages. Hospital biochemists and laboratory technicians may obtain some appreciation of the value and limitations of the chemical analyses they are called upon to perform.

Diabetes, liver function and to a lesser extent renal function have been considered in somewhat greater detail than the other subjects. In the case of diabetes no apology is needed, for the biochemical changes occurring in diabetic coma provide a valuable synopsis of very nearly the whole of chemical pathology. Liver function has been allotted two chapters, partly in reflection of the complexity of the subject and partly because some introduction to the general aspects of liver function appeared necessary in order to render intelligible the results obtained when liver function tests are carried out in different diseases.

It is hoped that the research worker will not find too much in the account of his own subject with which to quarrel, for this textbook is not intended for him. The specialist on water and electrolytes may find a possibly over-simplified account of water deficiency, of sodium deficiency and of disorders of potassium balance. The author trusts that any errors may be of omission rather than of commission.

The few references after each chapter are chiefly included for the post-graduate, although some students may care to look more deeply into subjects which they find of particular interest.

It is a pleasure to acknowledge my thanks to Professor A. C.

Frazer for Figs. 13 and 14; to Dr. Fuller Albright, Dr. Gregory Pincus and the Academic Press Incorporated for their permission to construct Figs. 9, 10, 11 and 12, which are the simplified versions of four diagrams appearing in *Recent Progress in Hormone Research*; to Professor J. P. Peters, Dr. Garfield G. Duncan and the W. B. Saunders Company for their permission to reproduce Fig. 3; to Professor J. L. Gamble for the block diagrams shown in Figs. 1 and 4.

The following have kindly given me permission to quote from my own previously published articles: Messrs. Methuen, Ltd. (jaundice, including Fig. 5); The Oxford University Press and the editors of the *Quarterly Journal of Medicine* (liver-function tests, including Fig. 6); Dr. S. C. Dyke, the editor of *Recent Advances in Clinical Pathology*, and Messrs. J. & A. Churchill Ltd. (chemical changes in diabetic coma), and Sir Cecil P. G. Wakeley, the editor of Rose & Carless' *Surgery*, and Messrs. Baillière, Tindall & Cox (water balance, including Tables 1, 2 and 3).

Acknowledgements are due to Dr. M. H. Pond, Dr. M. J. H. Smith and Mr. J. Collings-Wells, who have provided valuable comments on some of the chapters; Professor R. Knox, who read through the whole of the manuscript; Dr. M. E. A. Powell, who so kindly read the galley proofs; Miss V. Newing for preparation of the typescript, and Miss M. Pickett for helping with the preparation of the figures.

C. H. G.

CONTENTS

Chapter		Page
	Preliminary Note on Units	1
I	Renal Function	5
II	Acid–Base Balance	19
III	Fluid Balance—Œdema	33
IV	Fluid Balance—Salt Deficiency and Water Deficiency	49
V	Liver Function and Disorders of the Liver	64
VI	The Plasma Proteins	81
VII	Calcium and Phosphorus	93
VIII	The Gastro-intestinal Tract	110
IX	Some Aspects of the Chemical Pathology of Diabetes	126
X	Biochemical Tests in Endocrine Disease	142
XI	Biochemical Genetics	160
XII	Clinical Enzymology	170
XIII	Some Aspects of the Chemical Pathology of the Nervous System	179
XIV	Some Biochemical Aspects of Hæmatology	187
XV	Some Miscellaneous Topics	201
	Appendix—Routine Tests	206
	Blood Analysis—Some Normal Values	222
	Index	227

PRELIMINARY NOTE ON UNITS

mEq, millimoles and m-osmols

THERE are many advantages in expressing the concentration of the plasma electrolytes not as g or mg/100 ml., but in terms of milligram-equivalents (mEq) of the ions per litre. The sum of the concentrations of anions must then equal the sum of the concentrations of the cations, a matter of particular convenience when the concentration or change in concentration of one ion needs to be compared with the concentration or change in concentration of another ion.

To calculate the concentrations in mEq/l., the concentration in mg/100 ml. is multiplied by 10 to give the concentration in mg/l. This is then converted into the concentration in mEq/l. by dividing the equivalent weight of the ion. This is illustrated by the following examples:

Thus 355 mg chloride/100 ml. $\equiv$ 3,550 mg/l.
Dividing by 35·5, the equivalent of chloride $\equiv$ 3,550/35·5
$\equiv$ 100 mEq/l.

325 mg sodium/100 ml. $\equiv$ 3,250 mg/l.
Dividing by 23, the equivalent of sodium $\equiv$ 3,250/23
$\equiv$ 141 mEq/l.

20 mg potassium/100 ml. $\equiv$ 200 mg/l.
Dividing by 39, the equivalent of potassium $\equiv$ 200/39.
$\equiv$ 5·1 mEq/l.

10 mg calcium/100 ml. $\equiv$ 100 mg/l.
Dividing by 20, the equivalent of calcium, i.e., half the atomic weight because it is divalent $\equiv$ 100/20
$\equiv$ 5 mEq/l.

The calculation of the bicarbonate concentration from the alkali reserve, however, requires special comment. The alkali reserve has usually been expressed as volumes CO_2 per cent and refers to the quantity of CO_2 liberated from the bicarbonate of the plasma by reacting with acid.

From the equation

$$NaHCO_3 + HA = NaA + H_2O + CO_2$$

1 equivalent of $\overline{HCO_3}$ forms 1 molecule of CO_2 (i.e., 1 g equivalent forms 1 g molecule of CO_2). Since 1 g molecule of a perfect gas occupies 22·4 l. at N.T.P., 1 g equivalent of bicarbonate ion will form 22·4 l. of CO_2, 1 mEq of HCO_3 therefore forms 22·4 ml. of CO_2.

The alkali reserve in volumes of CO_2 per cent is therefore multiplied by 10 to give ml. CO_2/l., and then divided by 22·4 to give mEq/l.

For many years it has been customary for clinicians and clinical pathologists to express concentrations of acid in gastric juice in apparently rather arbitrary units—namely ml. of $N/10$-HCl/100 ml. It does not appear to have been appreciated that these units are in fact mEq/l. Thus, if

$$\begin{aligned}\text{acid concentration} &\equiv x \text{ ml. } N/10 \text{ acid}/100 \text{ ml.}\\ &\equiv 10x \text{ ml. N}/10 \text{ acid}/l.\\ &\equiv x \text{ ml. } N \text{ acid}/l.\end{aligned}$$

But each ml. N acid by definition contains 1 mEq of acid ∴ acid concentration $= x$ mEq/l.

It is useful to represent graphically the ionic composition of the body-fluids as block diagrams in which the height of the sections of the blocks are proportional to the concentrations of the electrolytes. When these are expressed in mEq/l. the height of the block representing the pattern of the anions must be the same as that of the block representing the pattern of the cations. This is illustrated in Fig. 1, which shows the electrolyte pattern of six body-fluids represented in this way. Several points are worthy of comment.

1. The height of all the columns is the same, indicating the uniformity of total concentrations of ions in the body-fluids irrespective of the nature of these ions.

2. If the concentration of a cation is increased there must be either an equivalent decrease in another cation or else an equivalent increase in concentration of an anion. This last point is well shown by comparing the composition of a weakly acid gastric juice, in which only a small proportion of the cation column is occupied by hydrogen ions, with that of the highly acid gastric juice, in which most of the cation column is occupied by hydrogen ions with an equivalent reduction in the concentration of sodium ions. As will be seen later, if because acid gastric juice is vomited, there is a reduction of chloride ion in the extracellular fluid, there must also be an equivalent increase in the bicarbonate concentration.

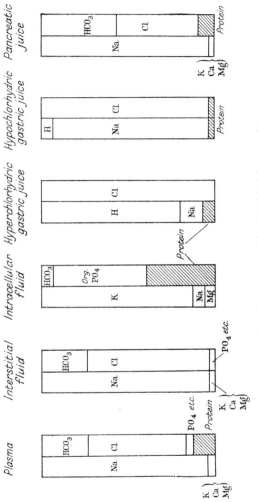

FIG. 1. Electrolyte pattern of body-fluids (after Gamble).

3. Protein appears as an anion at pH 7·3–7·5, the pH of plasma, and at the more alkaline reaction of pancreatic juice, but in acid gastric juice it appears as a cation.

Although concentrations of anions and cations are best expressed in mEq/l., these units cannot be used for expressing concentrations of substances such as urea, glucose and other organic compounds which do not ionize and are present in substantial quantities in the body fluids. Here the concentration of molecules is important, and the obvious unit would be the mole or millimole (i.e., the molecule weight expressed in g or mg) per litre. Concentrations are then expressed in terms of molarity or millimolarity. In the body fluids the osmotic pressures of substances are often of special significance, and this depends on the total concentrations of molecules and ions in solution. The concentrations of osmotically effective molecules or ions obviously cannot be expressed in terms of mEq or millimoles. The unit for expressing osmotic concentration is the osmol, which is the amount of substance in 1 l. of solution which under ideal conditions exerts an osmotic pressure of 22·4 atmospheres and will, therefore, depress the freezing point by 1·86° C. One g mole/l. of an un-ionized substance exerts an osmotic pressure of 22·4 atmospheres and is equivalent to 1 osmol/l. The osmotic concentration of a molar solution of an ionized substance will depend on the number of ions into which it dissociates, e.g., for sodium chloride it is 2 osmols/l. and for calcium chloride ($CaCl_2$) it is 3 osmols/l. For comparing osmotic effects (as opposed to chemical equivalence) the concentrations of both electrolytes and non-ionizing substances in the body fluids may be expressed in terms of osmolarity or osmolality. Osmolarity refers to concentration in osmols or milli-osmols per litre of solution; osmolality to concentration per kg of water. The latter is to be preferred with plasma and cells. Plasma and cells contain about 10 per cent or more of solids and red cells about 25 g of hæmoglobin per 100 ml.; this will result in considerable difference between osmolarity and osmolality. The osmolality of the body fluids is normally about 300 m-osmols/kg. water.

CHAPTER I

RENAL FUNCTION

Physiology of the kidney. The functional unit of the kidney, the nephron, of which there are about one million in each kidney, consists of a glomerulus with its tubule. The osmotic pressure in the cortical part of the kidney around the convoluted tubules is equal to that of the systemic plasma, but increases progressively in the medulla, reaching a maximum near the top of the papillae. About 1,200 ml. of blood pass through the normal kidney per minute and in the glomeruli are in contact with a filtering surface of about a square metre. The fluid passing through this filtering surface is normally an ultra-filtrate, i.e. it possesses the same chemical composition as the plasma except that it is relatively free of protein. About 120 ml./min. of this ultra-filtrate are formed and pass from the glomeruli into the tubules. Iso-osmotic absorption of electrolytes and water take place in the proximal convoluted tubule. Water is absorbed in the descending limb of the loop of Henle and electrolytes in the ascending limb, thus providing a counter-current multiplier and exchange system for maintaining the hyperosmolality in the inner part of the medulla and papillary tip. Only with such an osmolar gradient between cortex and papillary tip can highly concentrated urine, normally of volume about 1 ml./min., be formed from the fluid in the lumina of the distal and collecting tubules.

During passage of the glomerular filtrate through the proximal tubule 80 per cent of its content of sodium chloride and other ions and water are absorbed iso-osmotically, so that there passes from the end of the proximal tubule about 24 ml. (20 per cent of the glomerular filtration rate) of fluid of osmolality similar to that of normal interstitial fluid, i.e., about 300 m-osmol/l. Water is absorbed passively throughout the centimetre or so length of the descending limb of Henle, which passes into the medulla where the osmolality of the interstitial fluid steadily increases to about 2,100 m-osmol/l. About 6 ml./min. of tubular fluid now of concentration nearly 2,100 m-osmol/l. reaches the loop of Henle and on passing through the ascending limb equilibrates with interstitial fluid of gradually diminishing osmolality. No shift of water occurs, although sodium ion with chloride following passively is actively transferred from the tubular fluid into the interstitial fluid, so that about 6 ml./min. of fluid of 300 m-osmol/l. reach the beginning of the distal convoluted tubule. The ascending loop of Henle thus acts as a counter-current multiplier system requiring energy for its maintenance by absorption of sodium from the tubular fluid into the interstitial fluid. The high osmolality of the inner part of the medulla would be rapidly dissipated by the circulating blood if

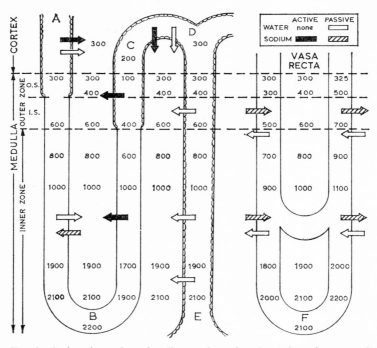

Fig. 2. Active absorption of sodium and passive absorption of water and sodium in the loop of Henle and in the vasa recta. The figures show the concentration in m-osmol/l. of the interstitial fluid, the fluid in the tubule lumen and the blood plasma in the vasa recta. The loop of Henle acts as a counter-current osmotic multiplier, the vasa recta act as counter-current exchangers (see text). (After Gottschalk & Mylle.)

the blood supply to the inner part of the medulla were not arranged as a counter-current exchange system, the blood in the descending arterial limb of the vasa recta exchanging with that in the ascending venous limb. Sodium chloride and urea pass from the returning venous limb into the arterial ascending limb, and the blood becomes steadily more and more hyperosmolar, until it is about 2,100 m-osmol at the papillary tip; the blood returning in the venous end returns to normal osmolality by the time it reaches the cortico-medullary junction. The blood supply through the vasa recta is slow compared with that through the cortex, and there is ample time for counter-current exchange to occur.

Only about a quarter of the loops of Henle pass into the hyperosmolar area of the medulla. The remainder are too short to act as counter-current multipliers and to permit the diminution in volume of the tubular fluid during its passage through the descending loop of Henle. There is thus presented to the distal tubules up to about 18 ml. of fluid/min., 6 ml. from the long loops of Henle and the remainder from the short loops.

The distal tubules and collecting tubules of all nephrons pass through the hyperosmotic zone to reach the papillary tip to discharge into the renal pelvis. In diabetes insipidus with absence of anti-diuretic hormone (ADH) the collecting tubules are impermeable to water, but active absorption of sodium ions (passively followed by chloride ions) continues and about 18 ml. hypotonic urine will pass out into the renal pelvis. In the presence of circulating ADH the collecting tubules become permeable to water, which is passively reabsorbed by the osmotic gradient between the tubular fluid and the renal tissue so that under maximal ADH action about 1 ml./min. of concentrated urine of 2,100 m-osmol/l., only slightly less than that of the interstitial fluid at the papillary tip, passes into the renal pelvis. Sodium absorption, with and without ion exchange (see below) in the distal tubules, is mediated by aldosterone and other mineralocortical hormones of the adrenal cortex. Direct reabsorption of sodium without ion exchange can occur only if chloride is available to pass passively with the sodium. If sodium is accompanied by a less diffusible anion, e.g., sulphate, bicarbonate or phosphate, a high electrochemical gradient is produced and impedes further direct absorption.

Although about 3 mEq bicarbonate/min. passes into the glomerular filtrate, the urine contains no bicarbonate and normally is acid. The reabsorption of bicarbonate occurs throughout the whole nephron. The reabsorbed sodium ions exchange for hydrogen ions from intracellular carbonic acid formed from carbon dioxide and water and facilitated by the action of carbonic anhydrase. Carbonic anhydrase inhibitors slow but do not inhibit the combination of carbon dioxide and water, so that bicarbonate reabsorption is not entirely abolished by such diuretics but leads to an alkaline urine containing appreciable amounts of unabsorbed bicarbonate. Bicarbonate excretion is profoundly influenced by changes in acid–base balance (see p. 19).

Potassium is probably completely reabsorbed in the proximal tubule, but is excreted with hydrogen ions entirely in the distal tubule by exchange with the reabsorbed sodium; potassium and hydrogen ions compete with each other for the available transport systems. Potassium excretion is thus increased by reduction of hydrogen-ion exchange due to carbonic anhydrase inhibitors, as well as by administration of potassium salts leading to an alkaline urine. Excess potassium loss occurs: (i) if sodium is available for ion exchange, since if all the filtered sodium has been reabsorbed in the proximal part of the nephron none will be available for ion exchange with potassium; (ii) if ion exchange is exaggerated when the sodium in the tubule is accompanied by a non-diffusible anion such as bicarbonate rather than by chloride; and (iii) if ion exchange is exaggerated by excess circulating aldosterone. In renal tubular acidosis (see p. 26) the deficiency of hydrogen ions in the urine results in an increased loss of potassium. This may occur in sodium deficiency after continued use of diuretics or in refractory œdema associated with cardiac failure, nephrotic syndrome or cirrhosis of the liver.

Renal clearances. The clearance of a substance is defined as the maximum volume of blood which would be completely cleared of the substance per minute, i.e., the volume of blood required to supply that quantity of the substance excreted per minute in the urine.

e.g., if volume of urine excreted per minute $= V$ ml.
and if concentration of substance in urine $= U$ mg/ml.
then quantity of substance excreted per minute $= UV$ mg
and if concentration of substance in blood $= B$ mg/ml.
volume of blood providing UV mg of the substance (i.e., the clearance) $= \dfrac{UV}{B}$ ml.

Some substances are unequally distributed between the red cells and the plasma, and since plasma is more directly concerned with urine formation than red cells, we are more often concerned with the plasma clearance, i.e., the maximum volume of plasma required to supply that quantity of the substance excreted per minute.

On this definition the clearance has an actual numerical relation to a definite volume of blood or plasma only when considered in relation to a substance which is completely cleared from the blood or plasma passing through the kidneys. However, even with substances which are not completely cleared, the concept of clearance has been of the greatest practical value. Some substances, such as mannitol or inulin, are not completely cleared, but the determination of their clearance is believed to provide an actual measure of the rate of glomerular filtration. These two substances are believed to be neither excreted nor absorbed by the renal tubules, so that the quantity UV mg excreted per minute in the urine—i.e., that passing from the distal ends of the tubules—must equal that passing into the proximal ends of the tubules (i.e., the quantity produced per minute in the glomerular filtrate). The concentration in the glomerular filtrate is P, the concentration in the plasma, and UV mg must be transported in $\frac{UV}{P}$ ml. of glomerular filtrate, i.e., the glomerular filtration rate in ml./min. $= \frac{UV}{P} =$ plasma clearance. This relationship is valid only for substances which are neither excreted nor absorbed in the renal tubules. This may be true of inulin and mannitol in the normal kidney, but it is not necessarily so in the pathological kidney.

The clearance of many substances are much greater than those of inulin and mannitol because tubular excretion results in a more complete clearance of plasma than can be caused by glomerular filtration alone. Diodrast and *para*-aminohippuric acid (PAH) in low plasma concentrations have identical and high plasma clearances. By definition this is equal numerically to the maximum volume of plasma that may theoretically be completely cleared; if these substances are not completely cleared, the renal plasma flow must be higher still. None has yet been found with higher values, and if diodrast or PAH is completely cleared from the plasma by passage through the kidneys, as has been shown by renal catheterization to be reasonably true in normal man but not always in patients with renal disease, the numerical value of their clearances must be equal to the renal plasma flow.

The urea clearance is rather easily estimated, but is numerically less than the glomerular filtration rate because of diffusion of urea from the fluid in the tubules to the blood. The amount of urea so diffusing back will be greatest when the minute volume of urine is small, so that the urea clearance falls with diminishing urine flow. For urine flows greater than 2 ml./min. the urea clearance is sensibly constant and at a maximum value. The average **maximum clearance** of normal individuals is 75 ml./min., and values observed in clinical studies are usually expressed as a percentage of this value. However, patients do not always excrete urine at rates greater than 2 ml./min., and then the urea clearance of a patient cannot be compared with the average normal maximum clearance, but would need to be compared with the average normal clearance for the prevailing urine flow. Instead, however, the observed clearance is arbitrarily multiplied by $1/\sqrt{V}$ to give a figure (the so-called **standard urea clearance**) which is intended to represent what the patient's clearance would be if his urine flow were 1 ml./min. It is essential to realize that this standard clearance is not an actual volume of cleared blood, but is the result of a mathematical procedure. The average normal standard urea clearance is 54 ml./min., observed standard urea clearances always being expressed as a percentage of this value. Changes in the urea clearance are roughly parallel to changes in the glomerular filtration rate, but the determination does not provide a numerical measure of glomerular filtration.

The creatinine clearance is frequently measured to provide an estimate of the glomerular filtration rate, but since creatinine is excreted by the renal tubules, the values obtained are too high. Moreover, normal plasma concentrations of creatinine are too low for accurate measurement, and hence errors in the measurement of creatinine clearance are very great. The clearance of injected radioactive cobalamin vitamin B_{12} is readily measured and provides a convenient means of measuring glomerular filtration rate. In renal insufficiency when the high plasma concentration is more readily determined the glomerular filtration rate will be very low and of little value.

Maximum tubular excretory or reabsorptive capacity. The tubular excretion of a substance is given by the total amount excreted per minute in the urine less the amount produced by filtration. The amount filtered is equal to the product of the glomerular filtration rate (inulin clearance) and the concentration in the glomerular filtrate (concentration in plasma). Tubular excretion of PAH increases with plasma concentrations up to a certain critical level, above which tubular excretion is constant. The amount of PAH in mg excreted per minute by the tubules at plasma concentrations above the critical level is believed to represent the maximum tubular excretory capacity of the kidney for PAH (i.e. PAHT_m). Tubular reabsorption can be calculated from the amount of glucose passing per minute into the filtrate (the product of the inulin clearance and the concentration of plasma glucose) less the amount passing into the urine per minute. There is a maximum reabsorptive rate for glucose (glucose T_m) which cannot be exceeded despite increase in the plasma concentration.

The width of the renal medulla in different species corresponds with the dryness of their environment and with the highest attainable urine concentration. In some species peculiar to an arid environment the renal papilla is long and even projects into the lumen of the ureter. The ability of the kidney to produce a concentrated urine is therefore dependent upon the width of the renal medulla. Medullary function may be assessed by measurement of the free water absorption. This is given by the osmolar clearance less urine volume. The osmolar clearance is the volume of urine required to excrete the same amount of solute at an osmolar concentration equal to that of plasma, i.e.

$$\frac{\text{(Urinary osmolarity} \times \text{Minute volume)}}{\text{Plasma osmolarity}}$$

These investigations are not employed routinely, but have given insight into the mechanism of many abnormalities of renal excretion.

Proteinuria. Proteinuria may be in part due to an increased permeability of the glomerulus, to impaired reabsorption of normal amounts of protein in the glomerular filtrate or to tubular excretion. Of these the last two factors are probably the more important. Normal urine contains a small amount of protein insufficient to give positive reactions to the usual clinical tests, the sensitivity of which unfortunately depends on the composition of the mixture of proteins excreted. The normal daily amount is about 20–40 mg., and an increase is important in the diagnosis of renal disease, and especially of the early stages of pyelonephritis.

Transitory proteinuria occurs in fever, physical exertion and even emotion and never indicates renal disease. Orthostatic benign proteinuria occurring in the upright position occurs in 77 per cent of all normal adolescents of 14–16 years; the maximum protein concentration is usually less than 3 g./100 ml.

Hæmoglobinuria is a form of proteinuria and may occur without renal damage in hæmolytic states. Otherwise continuous proteinuria is indicative of renal damage, even though the severity of proteinuria may increase with the upright position. Red cells and casts must then be looked for.

The protein excreted in the nephrotic syndrome is mainly albumin, but increased amounts of globulin are found in other renal lesions. Using immunoassay methods or gel filtration, the renal clearances of proteins of greatly differing molecular size, such as α_2 macroglobulin, transferrin and albumin, may be measured and are said to be important in diagnosis and in assessing the prognosis of the nephrotic syndrome responding to steroids.

Impairment of renal function. In renal impairment glomerular and tubular function are usually simultaneously impaired, but in the early stages of acute nephritis glomerular damage often precedes the tubular damage by a few hours.

Impairment of renal function may be classified into *renal insufficiency*, in which the concentrations of the end products of metabolism in the body-fluid are maintained within normal limits, and *renal failure*, in which the end products accumulate in excess.

Renal insufficiency. In this condition the composition of the blood is maintained within normal limits in all ordinary circumstances, yet special tests of renal function reveal an inability to meet any additional demands. During this stage of renal impairment the clearances may be decreased to 40 per cent or less of normal. There is an inability to excrete dilute urine with specific gravity approaching 1·002 under conditions of excessive hydration, as well as an inability to excrete concentrated urine of specific gravity of nearly 1·030 when there is water deprivation.

With increasing severity of the impairment of renal function the specific gravity becomes increasingly fixed within the limits 1·010–1·012 isosthenuria. The concentrations of end products of metabolism in the urine are not so high as is normal, and large daily volumes of urine are necessary to maintain the composition of the blood within normal limits. Nevertheless, the urinary elimination of water is impaired, for in the simple water-elimination test, 1 litre is not excreted within the customary 4 hours, even though 4 or 5 litres of water per day need to be ingested and excreted in the urine to eliminate the end products of metabolism. In accord with the definition of renal insufficiency, blood urea, creatinine, uric acid and other contributors to the non-protein nitrogen, as well as the plasma inorganic phosphate, are within normal limits. In the later stages of impairment of renal function these substances accumulate in excessive amounts in the blood, and the condition merges into renal failure.

Renal failure. In acute renal failure, as observed in acute glomerulo-nephritis or anuria and in less acute form in malignant hypertension, the rate of excretion of the nitrogenous end products of metabolism is much less than their rate of formation in the body, and there is a rapid increase in their concentration in the blood (i.e., there is azotæmia). If the original cause of the renal failure is not successfully treated, death will rapidly take place.

In chronic renal failure—e.g., in chronic nephritis or other forms of slowly progressive destruction of the kidney (tuberculous kidney, polycystic kidney, tumours of the kidney, etc.)—the azotæmia is practically constant over long periods of time. The term "uræmia" is often used clinically in preference to "azotæmia", but, since uræmia is almost always associated with increases in nitrogeneous constituents of the blood other than urea, the term "azotæmia" is to be preferred.

Considering the excretion of urea, and ignoring the effect of the tubular reabsorption of this substance, the quantity excreted in the urine per minute must depend on the quantity of urea passing per minute into the glomerular filtrate. The amount of urea filtered per minute at the glomerulus is equal to the product of the glomerular filtration rate and the concentration of urea in the plasma. In chronic renal failure the glomerular filtration is greatly reduced, but excretion is kept at a normal level by an increase in the concentration in the plasma. The reduced glomerular filtration leads to an increase in the concentration of end products of metabolism in the blood, which results in a balance of excretion and production. Thus when the process leading to destruction of nephrons is stationary or is very slowly progressive, the concentration in the blood of end products of metabolism may be maintained at constant but elevated levels over long periods of time. In azotæmia the concentrations of non-protein nitrogen, urea, creatinine, uric acid and inorganic phosphorus in the blood are raised and the accumulation of acidic products reduces the plasma bicarbonate. The determination of the concentrations of these substances in the blood may be of value in assessing the severity of renal failure, and if repeatedly performed are valuable aids to prognosis. The specific gravity tests and the clearance tests all show gross impairment of renal function, and are unnecessary in renal failure.

The increase in the blood urea may not always be directly related to the severity of renal failure, for the exact level of the blood urea will depend not only on the state of the kidneys, but also on the protein intake and on the presence or absence of fever. The blood creatinine is more consistently related to the severity of renal failure, but the determination of normal concentrations is difficult.

Types of azotæmia. Azotæmia may be: (1) renal, or (2) extra-renal.

In renal azotæmia renal failure is due to disease of the renal tissue itself. In extra-renal azotæmia the renal failure is primarily due either to obstruction (post-renal azotæmia) or to some abnormality of the circulation (pre-renal azotæmia). As soon as the obstruction or the circulatory abnormality is successfully treated, renal function returns to normal, unless the condition has been present for so long that permanent renal damage has occurred. Pre-renal azotæmia occurs in conditions associated with low blood pressure and/or a decrease in plasma volume either absolute or, as in shock, relative to the size of the vascular bed. It occurs in cardiac failure, where deficient oxygenation and impeded circulation rapidly result in renal failure, as well as in the salt deficiency of severe sweating, vomiting, diarrhœa, intestinal obstruction and Addison's disease. In salt deficiency the loss of electrolyte-containing fluids leads to a reduction in the extracellular fluid volume, and the consequent diminution in the volume of the circulating plasma is intimately related to the subsequent renal failure (see p. 52). Intestinal hæmorrhage also causes a marked azotæmia, and here not only does the condition develop because of impaired renal function consequent to the altered dynamics of the circulation, but the digestion and subsequent metabolism of large amounts of hæmoglobin and plasma protein provide increased amounts of end products of metabolism which require excretion. Trauma and shock are also associated with pre-renal azotæmia due partly to impairment of renal function and partly to increased demands for excretion brought about by increased protein breakdown (hypercatabolism).

Post-renal azotæmia may be due to either partial or complete obstruction of the renal tract by a stone, by an enlarged prostate, by urethral stricture or by a congenital abnormality. The mechanism whereby a partial obstruction leads to impairment of renal function and subsequent renal failure is not yet clear, but distortion of the medullary papilla may be responsible (see p. 11).

Anuria and oliguria. After severe hæmorrhage, especially ante- or post-partum, severe toxæmia such as from *C. welchii* infection, crushing injuries, mis-matched transfusions and major trauma, sometimes after major surgery, and after sulphonamide administration with deficiency of fluid intake, anuria or, more usually, severe oliguria, may occur. Such anurias and oligurias are often due in the first place to drastic reduction in glomerular filtration, and the older forms of treatment involving intravenous administration of sucrose, hypertonic sulphate, etc., are irrational, for not only do they disturb water and electrolyte relationships in the body fluids but they act merely by reducing tubular reabsorption of water; any diuresis which may follow must be little more efficient in excreting the end products of metabolism than the minute amounts of urine which are usually produced during this condition. After correction of any anæmia by administration of compatible whole blood, the fluid intake should be just sufficient to balance the inevitable water loss due to losses of water in the expired air and in the perspiration. In temperate climates and in the absence of excessive sweating this should amount to rather less than 1 litre per day. A high-calorie, low-protein diet such as the Borst diet, consisting of a suspension of fat in glucose solution, may be given, so that the demands for excretion of nitrogenous end products are minimized. This diet is very unpalatable, and it is preferable to administer the fluid intake as 40 per cent glucose in water as an intra-caval drip. Such cases, conservatively treated in this way, often respond with a spontaneous diuresis and ultimate recovery. When diuresis begins, the fluid intake should be increased to balance not only the inevitable water loss, but the urinary volume as well.

Hæmodialysis. The availability of hæmodialysis apparatus (artificial kidney) has profoundly altered the treatment and prognosis of patients with oliguria and anuria. There are two forms of this apparatus, one with a large dialysing coil that can be sterilized but requires a large amount of blood to fill it and the other, the Kolff apparatus, with two smaller dialysing coils renewable after each dialysis and requiring much less blood. Each has its special advantages, and in certain circumstances when the apparatus is not available peritoneal dialysis can be employed. In this a suitable dialysis fluid is intermittently introduced and removed into the peritoneal cavity, the serosal surface of which functions as the dialysing membrane.

In all forms of dialysis the composition of the dialysing fluid will depend upon the nature of the material to be removed from the body; however, sometimes a higher concentration than usual of a constituent such as bicarbonate or potassium can be useful in correcting a deficiency. Such equipment has proved life-saving in patients with oliguria and anuria, and if used repeatedly can prolong the lives of patients with chronic renal failure. The apparatus is of greatest value in acute renal failure, but has been used for the elimination of poisons when the usual measures of forced diuresis by infusion of mannitol and bicarbonate have failed.

The nephrotic syndrome. This is characterized by œdema, hypoproteinæmia and massive proteinuria usually exceeding 5 g/day, often 10–15 g/day, but sometimes reaching 60 g/day. The condition may occur during the sub-acute stage of glomerulonephritis (Ellis Type II nephritis), amyloid disease, diabetes, polyarteritis nodosa, disseminated lupus erythematosus, chronic pyelonephritis, thrombosis of the renal veins and after certain drugs, including mercury-containing diuretics and teething powders.

The proteinuria may be theoretically due to increased glomerular permeability or to decreased capacity of the tubules to reabsorb protein. The first of these is more important and, because of the high rate of normal filtration, must be accompanied by increased tubular reabsorption of protein; this may well be responsible for the histological changes in the tubular cells. The unabsorbed proteins passing into the urine are therefore qualitatively the same as the plasma proteins; however, the albumin with its low molecular weight is excreted in greater amounts than the globulins, which nevertheless account for up to 30–40 per cent of the total proteinuria.

The plasma proteins show a decrease in all fractions except α_2-globulin which is increased. This hypoproteinæmia is due not only to loss of protein in the urine but also, as shown by studies with ^{131}I-labelled albumin, to increased catabolism of protein. Such studies have shown that protein synthesis may sometimes be increased, but is decreased in the most œdematous patients. Nevertheless, interpretation of the results of some studies of this kind has been made difficult by degradation of the albumin during introduction of the isotope label.

The blood lipids are usually raised, and the total cholesterol in the plasma is often increased from 400 to 600 mg/100 ml. or even to 1,000 mg/100 ml. This may be related to the raised α_2-globulin. The total calcium concentration is decreased on account of the decreased binding to plasma albumin, which may fall to below 1 g/100 ml. The ionized fraction of the calcium may also sometimes be low and be associated with tetany, especially in children. Plasma potassium may be low in association with a deficiency of potassium caused by the secondary aldosteronism often present, as well as to the dietary deficiency of potassium due to the illness of the patient. Renal function may be normal, as shown by measurement of renal plasma flow and glomerular filtration rate; indeed, urea and creatinine clearance may give higher values than normal and may cause the blood urea to fall to 10–15 mg/100 ml. However, renal function usually deteriorates gradually, proteinuria diminishes and œdema recedes.

A series of renal tubular defects are characterized by a failure of the renal tubules to absorb amino acids, glucose or phosphate.

The defect in tubular absorption may affect one, two or more of these substances, and in addition there may also be a defect of acidification of the urine in that the renal tubules are unable to absorb bicarbonate. This may lead to hyperchloræmic acidosis (see p. 26). Such defects in renal tubular function may be hereditary or acquired. Many of the former may be due to genetically determined enzyme defects (see Chapter XII), and most of the latter are secondary to some other hereditary defect, such as amino aciduria secondary to galactosæmia (see p. 166) and hepatolenticular degeneration (see p. 182). Some are due to defects of the proximal tubules, including renal glycosuria due to reduced glucose re-absorption, vitamin D resistant rickets or osteomalacia due to impaired absorption of phosphate. In the Fanconi syndrome, there is the complete array of renal tubular defects with impaired absorption of amino acids, glucose, phosphate, uric acid and bicarbonate. In cystinelysinuria there is impaired proximal tubular absorption not only of cystine and lysine but also of arginine and ornithine. In Hartnup disease many amino acids are inadequately reabsorbed, but not arginine, ornithine, glycine or proline.

Disorders of the distal tubule include renal tubular acidosis of infants, the renal tubular acidosis with nephrocalcinosis of adolescents and nephrogenic diabetes insipidus.

Further Reading

BLACK, D. A. K. 1965. "Renal Rete Mirabile." *Lancet*, 2, 1141.

BLACK, D. A. K. 1967. "Renal Disease." (2nd edition). Blackwell Scientific Publications: Oxford, Edinburgh.

DE WARDENER, H. E. 1967. "The Kidney—Outline and Normal and Abnormal Structure and Function." (3rd edition). Churchill, London.

CHAPTER II

ACID–BASE BALANCE

DISTURBANCES in acid–base balance consist of disproportion between acids and bases as defined later and are not primarily concerned with the proportions of anions to cations.

Although it is customary to classify disturbances of acid–base balance into alkalosis and acidosis, these terms are not always clearly defined. Alkalosis is used to describe a disturbance of acid–base balance which, if sufficiently severe, results in an increase of pH of the extracellular fluid to a value above 7·5—the upper limit of normality. Secondary changes by regulatory systems described below may prevent a rise in pH above 7·5. In this instance the alkalosis is said to be "compensated". When the pH does, in fact, rise above 7·5 the alkalosis is said to be "uncompensated" and the condition is better referred to as alkalæmia. Similarly, an acidosis is a disturbance of acid–base balance which, if sufficiently severe, results in a decrease in pH of the extracellular fluid to below 7·3—the lower limit of normality. If secondary changes prevent the fall in pH below this limit the acidosis is "compensated", but if, despite the secondary changes, the pH does fall below 7·3 the condition is better referred to as acidæmia.

Disturbances of acid–base balance are frequently accompanied by more general disturbances of Na^+, K^+ and Ca^{++} ions. Thus bicarbonate cannot be lost by diarrhœa or by urinary excretion except when accompanied by a cation, usually Na^+ and to a lesser extent by K^+. Similarly, H^+ will be lost in vomit accompanied by Cl^-. The pH of the body fluids is maintained within very close limits by buffers, by changes in respiration and by changes in renal excretion. These three mechanisms will be considered separately, although in fact they are closely integrated.

Buffer mechanisms. A solution is said to be buffered if upon the addition of small amounts of acid or alkali the pH does not alter significantly. The amount of acid or alkali required to produce a significant change in pH, under standard conditions, is a measure of the *buffer capacity* of the solution. The greater the amount of acid or alkali required to produce such a change, the better the solution acts as a buffer. Such buffer solutions contain mixtures of weak acids and their salts or weak bases and their salts. The former type accounts for virtually all the buffering activity of extracellular fluid. The latter type is undoubtedly of importance in intracellular buffering.

In modern terminology acids are compounds or ions capable of donating hydrogen ions; anions are known as the *conjugate bases** of these acids because they can accept a hydrogen ion. Anions which are effective components of buffer systems are known as *buffer bases*. An alkali is a compound producing hydroxyl ions, e.g., sodium hydroxide, ammonium hydroxide or a substituted ammonium hydroxide such as choline.

The pH of a buffer solution between the approximate limits of pH of 4 and 10 is given by the equation

$$\text{pH} = \text{Constant} + \log_{10} \frac{\text{Molar conc. of conjugate base}}{\text{Molar concentration of acid}} \quad . \quad (1)$$

where constant $= \text{pK}, = -\log_{10}$ apparent dissociation constant of the acid.

Table 1 indicates the more important buffer systems of the body.

TABLE 1

Fluid compartment.	Acid.	Conjugate base.	
Plasma and interstitial fluid	H_2CO_3	$\overline{HCO_3}$	
	$H_2\overline{PO_4}$	$\overline{HPO_4}$	
Plasma	HPr	$\overline{Pr}$	Pr is plasma protein
Red cell	HHb	$\overline{Hb}$	
	$HHbCO_2$	$Hb\overline{CO_2}$	
Intracellular fluid †	HPr_t	$\overline{Pr_t}$	When Pr_t is tissue protein
	HP_{org}	$\overline{P}_{org}$	When P_{org} represents organic phosphates
Urine	$H_2\overline{PO_4}$	$\overline{HPO_4}$	
	H_2CO_3	$\overline{HCO_3}$	

* The old terminology whereby cations such as Na^+, K^+, Ca^{++}, etc., are called bases is wrong and must be abandoned.

† In intracellular fluid, buffer base/conjugate acid systems are important, e.g., buffering by anserine and carnosine in muscle cells.

These acids and their conjugate bases act as buffers because the acids can give hydrogen ions to neutralize any bases (e.g., hydroxyl or bicarbonate ions) which may invade the body fluids, while the hydrogen ions of any invading strong acid will react with the base of the buffer system to form the undissociated acid.

Thus
$$HA + \overline{OH} = \overline{A} + H_2O$$
$$\underset{\substack{\text{buffer} \\ \text{acid}}}{HA} + \underset{\substack{\text{invading} \\ \text{base}}}{H\overline{CO_3}} = \overline{A} + H_2CO_3$$
$$\underset{\substack{\text{conjugate} \\ \text{base}}}{\overline{A}} + \underset{\substack{\text{invading} \\ \text{acid}}}{\overset{+}{H} + \overline{Cl}} = HA + \overline{Cl}$$

When a mixture of buffers is present in the same solution the relative efficacies of the buffers depend upon their concentrations and upon the proximity of the pH of the solution to the pK_a values of the weak acid components of the buffers. Acids and their conjugate bases act most effectively as buffers when the pH is near the pK of the acid in question.

Equation (1) can be applied to *each* buffer species independently. Thus:

$$pH = pK' + \log\frac{H\overline{CO_3}}{H_2CO_3} = pK'' + \log\frac{H\overline{PO_4}}{H_2PO_4'} = pK''' + \log\frac{\overline{Prot}}{HProt}$$

where $pK' = pKa_1 = -\log_{10}$ (the apparent dissociation constant of H_2CO_3);

$pK'' = pKa_2 = -\log_{10}$ (the apparent dissociation constant of $H_2\overline{PO_4}$);

$pK''' = pKa = -\log_{10}$ (the apparent dissociation constant of HProt).

In a typical normal plasma the amounts might be $H\overline{CO_3} = 30$, $H\overline{PO_4} = 0.3$ and $Prot' = 1.7$ mEq/l. Although the pK'_{a2} of $\overline{H_2PO_4}$ is nearer to the pH of plasma than the pK'_{a1} of H_2CO_3, the carbonic acid–bicarbonate system is much the more important buffer because of its higher concentration and, as mentioned below, the volatility of its acid component.

The pH of blood plasma (and extracellular fluid) is given by

$$pH = pK' + \log\frac{H\overline{CO_3}}{H_2CO_3} \quad . \quad . \quad . \quad . \quad (2)$$

where HCO_3' = molar concentration of bicarbonate ion;

H_2CO_3 = molar concentration of carbonic acid (H_2CO_3 including dissolved CO_2).

In aqueous solution $pK' = 6.10$, the pK of the acid, but its value depends upon temperature and pH, and in a complex system, such as plasma or whole blood, can vary between 6.09 and 6.12.

For many years bicarbonate concentration in plasma has been and still may be measured volumetrically. The plasma was equilibrated with air containing a known partial pressure of CO_2

(usually normal alveolar air from the person carrying out the determination), and subsequently the volume of CO_2 liberated by acid was measured. After correction for the dissolved carbon dioxide (which could be calculated from the solubility and partial pressure of CO_2 with which the plasma was equilibrated) the bicarbonate concentration was obtained in vols. CO_2 at N.T.P./100 vols. plasma. This was the so-called "alkali reserve", it would be better called "the base reserve". In calculating pH according to equation (2), the carbonic acid concentration (i.e., concentration of physically dissolved CO_2) would also need to be expressed in vols. CO_2/100 vols. plasma. However, plasma bicarbonate is now always expressed as milli-equivalents per litre, and the concentration of carbonic acid must be similarly expressed. The concentration of physically dissolved carbon dioxide in arterial blood is related to the partial pressure, which is equal to that in the alveolar air, and for clinical purposes therefore carbonic acid concentration is still usually expressed in terms of its partial pressure, P_{CO_2} in mm Hg, and these units will be used subsequently.

The HCO_3/H_2CO_3 system, although an excellent buffering system, is, however, effective only for acids stronger than carbonic acid. Buffering of excess H_2CO_3 is carried out by the hæmoglobin–hæmoglobinate system within the red cells, and this buffer activity produces consequential ion movements known as the chloride shift. Carbonic acid passes into the red cells and reacts with the potassium hæmoglobinate according to the equation

$$H_2CO_3 \rightleftharpoons \overset{+}{H} + \overset{-}{HCO_3}$$
$\overset{-}{Hb}$ (hæmoglobinate ion)
$\updownarrow$
HHb (undissociated hæmoglobin)

The bicarbonate ion passes out of the red cell, and chloride ion passes from the plasma into the red cell. In this way any increase in plasma P_{CO_2} (H_2CO_3 concentration) is accompanied by a corresponding increase in bicarbonate concentration. The pH is therefore relatively unchanged, and the system acts as a buffer. Conversely, when H_2CO_3 is lost by the body the reverse change occurs, chloride passing out of the red cell, bicarbonate passing in and reacting with hæmoglobin to give the potassium hæmoglobinate and H_2CO_3. In this way the pH remains constant because the ratio of bicarbonate to P_{CO_2} is maintained virtually constant unless the body gains or loses too much H_2CO_3.

Respiratory mechanism. In the circulating blood the respiratory mechanism plays an important part in regulation of acid–base balance. If invasion by an acid should lead to diminution in the bicarbonate concentration, increased respiration (air hunger) leads to a reduction in P_{CO_2}, and again the fraction HCO_3/H_2CO_3 remains constant and the pH is unchanged. Conversely, respiration is depressed and the P_{CO_2} rises when invasion by a base leads to an increase in the bicarbonate concentration.

Renal mechanism. The buffers of the body fluids, the respiratory mechanism and the extra base which can be mobilized from the cells of the tissues are very rapidly available, and form the first line of defence against invasion by acid and alkali, but people who have once had a disturbance of acid–base balance do not possess permanently abnormal plasma bicarbonate concentrations. The more permanent correction of the acid–base balance of the body is effected by the kidney. Thus the invading acid (whether or not this is carbonic acid) is excreted in the urine, or if there is an excess of base, bicarbonate ions (together with sodium ions) will be excreted.

Intracellular pH. The intracellular fluid is slightly less alkaline than the extracellular fluid. Changes in the pH of the intracellular fluid parallel those of the extracellular fluid and occur rapidly on respiratory disturbance because of the free diffusibility of carbon dioxide and carbonic acid. In non-respiratory disturbance the changes are slow because bicarbonate is a relatively non-diffusible ion.

Classification

Disturbances of acid–base balance may be respiratory or non-respiratory (metabolic) in origin. In the respiratory disturbance there is a primary accumulation or loss of CO_2. On the other hand, in the non-respiratory metabolic type there is either a primary excess or a primary deficit of the bicarbonate concentration. Disturbances of acid–base balance may therefore be classified as follows:

1. Respiratory—
    (a) Primary CO_2 deficit.
    (b) Primary CO_2 excess.
2. Non-respiratory (metabolic)—
    (a) Primary bicarbonate deficit.
    (b) Primary bicarbonate excess.

1. Respiratory disturbances of acid–base balance.

(a) *Primary CO_2 deficit.* Overbreathing may occur in salicylate and aspirin poisoning, because of overventilation in a mechanical respirator, and is not uncommon in coma associated with injury to the brain after road accidents; it may also be a manifestation of hysteria. There is a primary fall in the P_{CO_2}, and this, in the compensated stages, is accompanied by a secondary fall in the bicarbonate concentration. But later this fall in the bicarbonate concentration fails to parallel the fall in the P_{CO_2}, the pH increases and there is an uncompensated alkalosis or alkalæmia.

(b) *Primary CO_2 excess.* This condition may occur if there is obstruction of the respiratory passages by laryngospasm, accumulated bronchial secretion or by an inhaled foreign body, if respiration is limited by skeletal deformities or paralysis of the respiratory muscles, if there is uneven distribution of the respired air because of emphysema or other disease in the lung or if the inspired air contains excess CO_2, as may occur if a mechanical respirator or anæsthetic apparatus is faulty. Thoracic injuries and open heart surgery may cause abnormalities of blood flow through the lungs and also cause primary CO_2 excess, which may also occur when the sensitivity of the respiratory centre is depressed by hypoxia or by drugs such as morphine, barbiturates or by anæsthetics. In all these conditions an increase of the gradient between CO_2 tension in the blood and that in the air is needed to maintain excretion of CO_2, which *must* always balance the production by the tissues. The disturbance may be precipitated or exaggerated by mild exertion or increased tissue metabolism caused by disease.

The retention of carbon dioxide will be minimized when possible by hyperventilation, which if inadequate will lead to an increased P_{CO_2}. The chloride shift will cause a secondary increase in the bicarbonate concentration, so that in the early stages the pH is not significantly changed, and the condition is compensated. At a later stage the pH will decrease significantly if the secondary increase in bicarbonate does not parallel the P_{CO_2}, and the disturbance is then uncompensated.

2. Non-respiratory (metabolic) disturbances of acid–base balance. These may be caused by

1. Intake of base, e.g., bicarbonate, or of acid, e.g., ammonium ion.
2. Loss of acid or base from the gastrointestinal tract by—
   (a) vomiting or aspiration of gastric contents (loss of hydrochloric acid);
   (b) diarrhœa causing loss of base (bicarbonate);
   (c) fistulæ causing loss of base (bicarbonate);
   (d) during therapy with ion exchange resins.
3. Metabolic disturbances—
   (a) ketosis in diabetes, starvation or toxic vomiting of pregnancy;
   (b) hyperchloræmic acidosis in surgical ureterocolostomy.
4. Primary disturbances of renal function in—
   (a) acute and chronic renal failure;
   (b) secondary to primary aldosteronism and Cushing's syndrome and during therapy with diuretics;
   (c) congenital tubular abnormalities, leading to a failure of the kidney to excrete hydrogen ions (i.e., hyperchloræmic acidosis).

(a) *Primary bicarbonate deficit.* Bicarbonate deficit may result from the invasion of the body by keto acids, due to starvation, vomiting of pregnancy or to diabetic coma, or in renal failure the kidney may retain the sulphuric acid and phosphoric acid produced by the metabolism of protein. Ingestion of calcium chloride or ammonium chloride leads to the virtual absorption of hydrochloric acid because the calcium is not absorbed and the ammonium ion is converted into urea. In these circumstances the acid reacts with the bicarbonate according to the equation

$$HA + NaHCO_3 = NaA + H_2CO_3$$

The plasma bicarbonate is therefore reduced. The respiratory centre is stimulated so that ventilation is increased and the P_{CO_2} falls. At first the diminution in bicarbonate and P_{CO_2} parallel one another and the pH, which depends on the ratio of these two concentrations (see equation (2)), is unaltered, but ultimately, if the disturbance of acid–base balance continues, the P_{CO_2} cannot be adjusted sufficiently to maintain the ratio constant, and the acid–base balance disturbance, so far compensated, becomes uncompensated.

Hyperchloræmic acidosis (with raised plasma chloride and diminished bicarbonate) occurs not only when alkaline intestinal secretions (containing the base bicarbonate) are lost but also in those hereditary defects of renal tubular function associated with failure of reabsorption of bicarbonate. Hyperchloræmic acidosis also occurs after transplantation of the ureters into the intestine for disease of the lower urinary tract. Studies with radioactive sodium and chloride have shown that the chloride ion present in the urine is absorbed from the intestine more rapidly than the sodium ion, perhaps as a result of its association with ammonia formed by bacterial breakdown of urea. Such hyperchloræmic acidosis is very effectively treated with large doses of alkalis and by ensuring rapid drainage from the bowel with a high fluid intake.

(*b*) *Primary bicarbonate excess.* If the body acquires an excess of bicarbonate, as may occur after ingestion of bicarbonate or after loss of highly acid gastric juice by vomiting or aspiration, ventilation is depressed, and in the early stages the increase in plasma bicarbonate is paralleled by a corresponding increase in P_{CO_2}. The disturbance is compensated, but ultimately the P_{CO_2} may not increase in proportion to the bicarbonate, and the ratio of the two will increase, the pH increases and the disturbance becomes uncompensated.

It is evident that unless it is known whether the primary disturbance is respiratory or not in origin, the determination of plasma bicarbonate alone does not indicate whether there is acidosis or alkalosis. Thus, a lowered plasma bicarbonate may be primarily metabolic or may be secondary to a primary respiratory CO_2 deficit (i.e., may be associated with either acidosis or alkalosis). Similarly, a raised plasma bicarbonate may be due to a primary metabolic bicarbonate excess or to a primary respiratory CO_2 excess (i.e., may be due either to a metabolic alkalosis or to a respiratory acidosis).

Changes in urine. Let us consider a non-respiratory acidosis in which the bicarbonate ion in the plasma has been reduced and replaced by the anion of the invading acid. This anion together with sodium ions will pass into the glomerular filtrate, and if exchanges of ions did not occur in the renal tubules the body would very rapidly become sodium deficient. However, the sodium ions are reabsorbed in the renal tubules and return to the bloodstream (see pp. 5–8). In their place hydrogen ions and ammonium ions pass into the tubular fluid, so that the fully formed urine contains the anions of the acid, with hydrogen ions and ammonium ions. The increase in hydrogen ions results in an increase in the titratable acidity, while the increase in ammonium ions leads to an increase in the so-called ammonia coefficient. This is the ammonia excretion expressed as a proportion of the total nitrogen excretion, and was originally used extensively as a measure of the severity of acidosis.

The urine contains large amounts of phosphates, which quantitatively form the most important buffer system in the urine. An excess of hydrogen ions causes an increase in the ratio of sodium dihydrogen phosphate to disodium hydrogen phosphate.

$$HA + Na_2HPO_4 \longrightarrow NaH_2PO_4 + NaA$$

It must be emphasized that this change in proportion is brought about by the presence of the acid itself in the urine. Similarly, in respiratory CO_2 excess the fluid in the renal tubules contains increased quantities of H_2CO_3 and $NaHCO_3$, the sodium of which, however, is replaced by hydrogen and ammonium ions. The urine therefore contains carbonic acid (i.e., increased P_{CO_2}) with ammonium and bicarbonate ions.

In non-respiratory bicarbonate excess the glomerular filtrate, like plasma, contains excess bicarbonate, which is excreted as such. The urine therefore becomes alkaline, and the phosphate present is converted to give a preponderance of the alkaline phosphate Na_2HPO_4.

$$NaHCO_3 + NaH_2PO_4 \longrightarrow Na_2HPO_4 + H_2CO_3$$

Two variations in this general pattern must be mentioned. In the non-respiratory bicarbonate deficit of renal failure due to primary kidney disease the replacement of sodium ions by ammonium ions is greatly restricted. Also in the late stages of non-respiratory bicarbonate excess brought about by loss of acid gastric juice in pyloric stenosis the overall deficit of sodium which is always present may become so great that all the bicarbonate with the associated sodium ions is reabsorbed by the renal tubules. The alkalosis and alkalæmia persist, yet the urine becomes acid.

Laboratory assessment of acid–base balance. In clinical work the respiratory or non-respiratory nature of the disturbance of acid–base balance is often well known from the nature of the disease. Then measurements of P_{CO_2} in respiratory disturbances or of bicarbonate concentration in non-respiratory disturbances respectively will be all that is required. Otherwise, for the precise investigation of acid–base balance two of the three parameters, pH, bicarbonate concentration or P_{CO_2} must be determined.

pH of arterial, capillary or venous blood arterialized by submerging the arm in hot water, can be determined with the requisite degree of accuracy with a number of commercial instruments. The blood must be collected anærobically and kept at 0° also anærobically until measurement, which should be performed as soon as possible.

The P_{CO_2} of arterial blood may be determined indirectly by measurements of CO_2 content of alveolar air, but this may not be practical in ill patients. For clinical purposes the P_{CO_2} of mixed venous blood can be determined by analysis of the CO_2 in a bag initially containing 0·3 per cent CO_2 after the subject has breathed and rebreathed several times. Some workers determine the CO_2 equilibrium content in a gas bubble introduced under the skin. The most accurate method depends upon the measurement of the pH of whole blood as collected and also after equilibration at two different partial pressures of carbon dioxide. Since the logarithm of the P_{CO_2} bears a straight line relationship to the pH, the P_{CO_2} of the blood sample as collected may readily be calculated. A special electrode is often used for direct electrometric measurement of P_{CO_2}.

Bicarbonate concentration is less easily measured than at first would seem the case. Most methods estimate physically dissolved CO_2 as well as the bicarbonate ion, and an appropriate and often arbitrary correction is often applied for the physically dissolved fraction. Table 2 shows the measures which various workers have adopted. Astrup's measurement of the "standard bicarbonate" has much to recommend it, since the bicarbonate is measured in the plasma after equilibration of *whole blood* with CO_2 at 40 mm. Hg. This results in the shift of chloride and bicarbonate ions across the red cell membranes, effectively reversing any secondary changes due to respiratory disturbance of acid–base balance. As seen in Table 3, the "standard bicarbonate", which provides a measure of the non-respiratory disturbance of

TABLE 2

Actual bicarbonate concentration (mEq/l.)	The bicarbonate concentration in plasma of anærobically drawn blood
Total CO_2 of plasma (mmol/l.)	The CO_2 derived from carbonic acid and bicarbonate in plasma from anærobically drawn blood
CO_2-combining power of plasma (mEq/l.)	The total CO_2 of plasma, separated at the actual P_{CO_2} from the cells, and equilibrated with CO_2 at a P_{CO_2} of 40 mm. Hg
Standard bicarbonate of Astrup (mEq/l.)	The bicarbonate concentration in the plasma separated from whole blood which has been equilibrated at a P_{CO_2} of 40 mm. Hg and with oxygen for full saturation of the hæmoglobin

acid–base balance, is independent of P_{CO_2} and oxygen saturation. However, at constant P_{CO_2} the carbonic acid–carbonate system accounts for only about three-quarters of the buffering power of

TABLE 3

	Hæmoglobin: P_{CO_2}, 20 mm. Hg.	Oxygenated: P_{CO_2}, 80 mm. Hg.	Hæmoglobin: P_{CO_2}, 20 mm. Hg.	Reduced: P_{CO_2}, 80 mm. Hg.
Total CO_2, mmol/l.	16·8	30·0	19·6	34·8
Plasma CO_2 combining power, mEq/l.	19·0	26·7	22·0	31·4
Standard bicarbonate, mEq/l.	21·2	21·2	21·2	21·2

Plasma values for total CO_2, CO_2-combining power and standard bicarbonate determined in samples from the same normal blood pool, at P_{CO_2} of 20 and 80 mm. Hg, respectively, and with the hæmoglobin completely oxygenated or completely reduced. The standard bicarbonate is independent of P_{CO_2} and oxygen saturation, thus showing its superiority to characterize the non-respiratory disturbances correctly.

(*By permission of Astrup*)

blood; plasma protein and the red cell hæmoglobin can also play a part as a result of the ionic interchange, and for this reason Singer and Hastings introduced the concept of buffer base, which is a measure of the total of the buffer anions, i.e., bicarbonate and protein ions. The buffer base normally amounts to 44–48 mEq/l. and is independent of P_{CO_2}. Its value and the extent of its deviation from normal may be read from a nomogram if P_{CO_2}, bicarbonate and pH of plasma are known.

A simpler assessment of the relative contributions of respiratory and non-respiratory factors to a disturbance of acid–base balance is obtained by plotting the observed pH and P_{CO_2} on a special diagram (the Davenport diagram, fig. 3(a)). Each of the

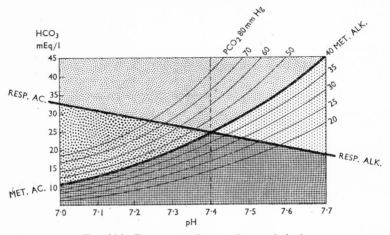

Fig. 3(a). Davenport diagram (hypermission).

(a) Respiratory alkalosis combined with metabolic acidosis.
(b) Respiratory acidosis combined with metabolic acidosis.
(c) Respiratory alkalosis combined with metabolic alkalosis.
(d) Respiratory acidosis combined with metabolic alkalosis.

curved lines is a P_{CO_2} isobar and represents the change in plasma HCO_3 with pH at constant P_{CO_2}. The straight line shows the normal change of HCO_3 and pH when there is uncomplicated respiratory acidosis or alkalosis. If there is primary loss or gain of HCO_3 due to non-respiratory acidosis or alkalosis the points representing observed pH and P_{CO_2} will be displaced to an area indicating a mixture of acid–base disturbances.

Astrup has devised a system of investigation of acid–base

balance based upon the measurement of the pH of whole blood as collected and after equilibration with 2 or 6 per cent carbon dioxide in oxygen. From the data obtained it is possible by a nomogram (fig. 3(*b*)) to determine P_{CO_2}, the "standard bicarbonate" concentration and the base deficit or excess, which provides a measure of the quantity in mEq of acid or base which has

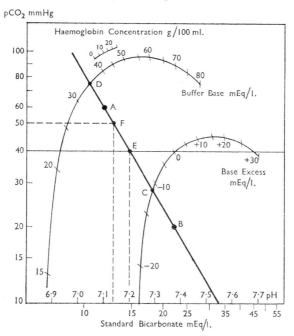

FIG. 3(*b*). Astrup diagram (by permission). Shows an example of the calculation of standard bicarbonate (point E), buffer base (point D), base excess (point C) and pCO_2 (point F) from three measured pH values, namely actual pH (point F) and pH after equilibration with two known CO_2-tensions (points A and B).

entered the bloodstream in order to bring about the observed change in buffer bases.

The precise investigation of acid–base balance by these means together with measurements of P_{O_2} and oxygen saturation is of the greatest importance in open-heart surgery with by-pass and hypothermia and is often of value in the investigation of cardiac disease or respiratory disease. Under these circumstances the state of acid–base balance is difficult to predict because of the complex effects of abnormal or artificial ventilation or circulation on P_{O_2} and P_{CO_2} and on the liberation of lactic, keto- and other acids into the body fluids.

NOTE ADDED IN PROOF

As considered above, the total hydrogen ion concentration is expressed as a negative logarithm, respiratory hydrogen ion as a partial pressure and non-respiratory hydrogen ion as standard bicarbonate in mEq/l. This is illogical and Whitehead and others have, therefore, recently proposed that all three factors should be expressed directly as hydrogen ion concentrations in nano-equivalent (nEq) which is 10^{-9} g. Eq. of hydrogen ion. The maintenance of a hydrogen ion concentration within a narrow range of normality is the ultimate concern of the organism, and this depends on respiratory and non-respiratory (i.e. metabolic) factors. Thus:

Total $[H^+]$ = Respiratory $[H^+]$ + Non-Respiratory $[H^+]$
if all $[H^+]$ are expressed in nEq/l.

The normal body hydrogen ion concentration expressed directly is 40 nEq/l (corresponding to a pH of 7·4) with a normal range of ±4 nEq/l. The respiratory and non-respiratory factors can be measured directly from a graph relating pCO_2 to hydrogen ion concentration (a straight line relationship when plotted on log/log paper). When the respiratory and non-respiratory hydrogen ion concentrations are known they may be plotted on a simple acid–base diagram, to allow the primary disturbance to be easily identified.

Further Reading

ANNOTATION. 1965. "Respiratory Failure." *Brit. med. J.* **1**, 1447.

ASTRUP, P., SIGGAARD ANDERSEN, O., JORGENSEN, K. and ENGEL, K. 1960. "The Acid-Base Metabolism. A new approach." *Lancet*, **1**, 1035.

CAMPBELL, E. J. M. 1965. "Respiratory Failure," *Brit. med. J.* **1**, 1451.

"Current Concepts of Acid-Base Measurement." 1966. *Ann. N.Y. Acad. Sci.* **133**. Art. 1.

DAVENPORT, H. W. 1958. *ABC of Acid Base Chemistry*. 4th edition. University of Chicago Press.

O'BRIEN, E. N. 1967. "Blood Gas and Acid Base Changes in Cardiac Surgery," *Bio-Medical Engineering*, **2**, 407.

SINGER, R. B. and HASTINGS, A. B. 1948. *Medicine (Baltimore)*, **27**, 223.

OWEN, J. A., DUDLEY, H. A. F. and MASTERTON, J. P. 1965. *Lancet*, **ii**, 660.

WHITEHEAD, T. P. 1967. "Blood hydrogen ion: Terminology, Physiology and Clinical applications." *Advances in Clinical Chemistry*, **9**, 195.

WYNN, V. 1965. "Water and Electrolyte Metabolism," in *Scientific Basis of Surgery*. Edited W. T. Irvine. J. & A. Churchill Ltd., London, p. 340.

CHAPTER III

FLUID BALANCE—ŒDEMA

SOME knowledge of the distribution of water within the body is essential for an appreciation of the mechanism of production, effects and rational treatment both of the various forms of œdema and of fluid deficiency, whether due to simple water deficiency or to loss of body fluids. The brief account which follows is in no way complete, but should provide the reader with the essential features of the subject.

Distribution of water in the body. The total body water amounts to between two-thirds and three-quarters of the total body weight. Part of this body water is within the cells and is called the intracellular fluid, and part is outside the cells and is called the extracellular fluid; the latter includes the water of the circulating plasma. That part of the extracellular fluid not included in the plasma is referred to as interstitial fluid. The synovial fluids of joints, peritoneal, pericardial, pleural, cerebrospinal and ocular fluids and lymph are also fractions of the extracellular fluids but are discontinuous with the interstitial fluid and plasma. From many points of view it is useful to regard the gastro-intestinal secretions also as part of the extracellular compartment.

The methods of determination of the volume of the body water and of that of its components are open to criticism and not always readily applicable to the sick patient. A measured amount of antipyrin, urea or tritiated water administered to an individual is believed, after 2 or 3 hours, to be evenly distributed throughout all the body water, and if the concentration in the body—i.e., in the blood—at this time is measured, calculation will provide a measure of the total body water. Extracellular fluid volume may be measured by the administration of measured amounts of thiocyanate, thiosulphate or radioactive sodium. These substances are believed not to pass into the cells and to be distributed uniformly throughout the extracellular fluid. A determination of the concentration of the substance in the plasma (itself a part of the extracellular fluid) will permit calculation of the volume through which it has been distributed (i.e., the volume of the extracellular fluid). Similarly, the plasma volume may be determined by the intravenous injection of the dye Evans blue or ^{131}I-labelled human albumin, which are considered to be distributed only in the plasma and not in the interstitial fluid. Determination of the intracellular fluid volume is much more difficult and requires extremely complex and indirect methods. Those in clinical use are usually only of value in measuring the *changes* in volume of intracellular fluid.

A rough calculation of the magnitude of the various fluid compartments from the body weight is often most useful and of the greatest importance in the treatment of fluid or salt loss in clinical practice. The quantities of water in the intracellular and extracellular compartments and in the plasma amount to about 50, 20 and 5 per cent of the body weight respectively. Table 4 shows the

TABLE 4—CALCULATED FLUID DISTRIBUTION IN FOUR INDIVIDUALS

	Weight.	Extracellular fluid.		Intra-cellular fluid.	H_2O.
		Plasma.	Interstitial fluid.		
Adult	11 stones = 70 kilos	3·5 litres	10·5 litres	36 litres	50 litres
,,	9 ,, = 58 ,,	2·9 ,,	9 ,,	28 ,,	40 ,,
Child	4½ ,, = 29 ,,	1·5 ,,	4·5 ,,	14 ,,	20 ,,
Infant	8 pounds = 3·6 ,,	180 ml.	540 ml.	2,280 ml.	3 ,,

TOTAL WATER OF BODY = 66–75 per cent body weight.
(Upper level in thin, lower level in fat persons, 80 per cent in infants.)
EXTRACELLULAR FLUID = approx. 20 per cent body weight.
(This includes plasma which amounts to approx. 5 per cent body weight.)
INTRACELLULAR FLUID = approx. 50 per cent body weight.

magnitude, calculated in this way, of the distribution of fluid in four individuals of differing size.

Although the nature of the ions and molecules is different in each compartment, the osmotic pressure is uniform throughout the body fluids, except where local activities are responsible for osmotic gradients e.g., in the renal medulla (see p. 6) and as seen below throughout the length of the capillaries. These disappear when the activity stops. The chief cation of plasma and interstitial fluid is sodium, with small, but important, amounts of potassium, calcium and magnesium. In intracellular fluid the chief cation is potassium. Chloride and bicarbonate are the main anions of interstitial fluid and plasma, but the latter also contains significant and very important amounts of protein which are not present in the interstitial fluid. On the other hand chloride and bicarbonate are low in the intracellular fluid, the principal anions of which are proteins, organic phosphates and organic acids. Many of these are polyvalent, so that it is difficult to equate ionic concentration with osmotic activity.

The concentration of potassium within the cell is maintained at 20–30 times the concentration in the extracellular fluid, and at the same time the concentration of sodium in the extracellular fluid is maintained at a much higher level than that in the intracellular fluid. This mechanism requires the expenditure of energy. In the red cell the energy required to maintain this relative exclusion of sodium from within the cells and potassium from the plasma is derived from the metabolism of glucose. If this glucose breakdown is inhibited potassium immediately diffuses out from, and sodium into, the cell. A similar mechanism is probably concerned in maintaining this effect in other cells. Apart from certain circumstances to be considered later, potassium and sodium may therefore be regarded as non-diffusible ions as far as the cell membranes separating the extracellular and intracellular compartments are concerned.

The distribution of fluid in the three main fluid compartments depends on the quantities and concentrations of non-diffusible colloids and ions (or effectively non-diffusible ions) in these compartments. Thus, the absolute volume of the intracellular and extracellular compartments depend on the absolute amounts of potassium and sodium in the body. The water contents of both compartments are regulated so that the osmolar concentrations are 285–310 m-osmol/l. In the extracellular fluid about 90 per cent of the osmotically active ions consist of the monovalent sodium, chloride and bicarbonate ions. Since the concentrations of polyvalent anions and cations are relatively small, the concentration of sodium ions is roughly equivalent to the sum of the chloride and bicarbonate ions, and the volume of extracellular fluid is determined by the quantity of sodium in the body. In the intracellular fluid the main cation is potassium, the concentration of which is maintained such that with the polyvalent anions, organic phosphate and protein, the osmolar concentration of which cannot readily be calculated from the ionic concentration, the osmolality is maintained equal to that of the extracellular fluid. Thus, the quantity of potassium in the body controls the volume of the intracellular fluid. When sodium ions are lost by the body, as when gastro-intestinal secretions are lost (see p. 51), the extracellular fluid volume will fall, potassium may pass out of the cell and be excreted, leading to a fall in intracellular fluid volume as well.

The active extrusion of sodium from the cells and passive retention of potassium within the cells is influenced by aldosterone as well as by other factors influencing the concentration of sodium and potassium in the extracellular fluid, so that under the influence of aldosterone and mineralocortical hormones sodium may enter and potassium leave the cells and be excreted in the urine.

Although the total amount of extracellular fluid is determined by the total quantity of sodium in the body, it is also influenced profoundly by the fluid balance between the plasma and the interstitial fluid.

The exchange of fluid between plasma and interstitial fluid takes place almost entirely in the capillaries, and the main factors controlling the passage of water and salts across the walls of the capillaries are:

1. Hydrostatic pressure of the blood in the capillaries.
2. Capillary permeability.
3. The difference between the osmotic pressure of the blood plasma and that of interstitial fluid.
4. Lymphatic drainage.
5. Tissue tension.

1. *Hydrostatic pressure of the blood in the capillaries.* Landis has shown that in the capillaries at the base of the nail-bed the hydrostatic pressures at the arteriolar end, in the middle and at the venous end of the capillary are 32 mm., 22 mm. and 12 mm. mercury respectively. This hydrostatic pressure results in a filtration of fluids through the walls of the capillaries.

2. *Capillary permeability.* The capillary is completely permeable to all crystalloids and water, but is normally impermeable to colloids such as plasma proteins. On the other hand, the

permeability to colloids may be increased by a number of agents such as certain poisons, oxygen lack and bacterial toxins.

3. *The difference between the osmotic pressure of the blood plasma and that of interstitial fluid.* Because of the specific permeability of the capillaries, the interstitial fluid is an ultra-filtrate of the blood plasma. In contrast to the plasma, which contains 6–8 g of protein per 100 ml., the tissue-fluid protein amounts to less than 0·1 g/100 ml. Apart from the effect of minor differences of salt concentration due to the setting up of a Donnan equilibrium, the osmotic pressure of the plasma and interstitial fluid will differ only by the osmotic pressure of the plasma proteins. At the arteriolar end this amounts to about 22 mm. of mercury. The plasma albumin will be more important than the plasma globulin as a contributor to the osmotic pressure of the plasma for the following reasons. The former has a smaller molecular weight, so that there are more molecules per g of albumin than molecules per g of globulin in plasma. Moreover, albumin is more dissociated into its ions because its iso-electric point (4·7) is further removed from the pH of normal plasma (7·4) than the iso-electric point of globulin (5·4). It is evident that in clinical problems in which the plasma osmotic pressure is of interest, consideration of the concentration of plasma albumin is of much greater importance than consideration of the albumin/globulin ratio, upon which many clinicians have laid great emphasis.

4. *Lymphatic drainage.* The interstitial fluid is drained by the lymphatic system. Some idea of the quantities of fluid removed from the interstitial fluid by the lymphatics has been provided by the work of Drinker, who has shown that after acute plasma protein depletion in the dog by plasmapheresis the lymph flow through the thoracic duct may be increased by as much as 200 ml./hour. Drainage of interstitial fluid via the lymphatics may become of major importance only when there is some disturbance of the water balance between plasma and interstitial fluid (see below).

5. *Tissue tension.* In some parts of the body, such as the palms of the hands and in the muscles, the interstitial fluid is less able to expand its volume because of the spatial arrangement of its connective tissues. On the other hand, the loose texture of the connective tissue below the eyes and behind the wrists resists much less readily any increase in volume of the extracellular fluid.

Fig. 4. shows how these factors are concerned in the dynamic equilibrium controlling the relative volumes of fluid in the plasma and in the interstitial fluid space. The hydrostatic pressure tends to produce an ultra-filtration of fluid from the plasma to the interstitial fluid. This hydrostatic pressure is greater at the arteriolar end than at the venous end. On the other hand, the effect of the plasma proteins is to withdraw fluid from the more dilute interstitial fluid into the more highly concentrated plasma. At the arteriolar end the hydrostatic pressure is greater than the osmotic pressure, and water and salts pass

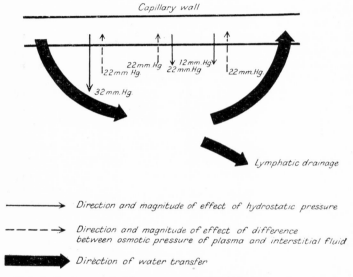

FIG. 4. Factors controlling movement of fluid between plasma and interstitial fluid.

from the plasma to the interstitial fluid. At the venous end the hydrostatic pressure is less than the osmotic pressure, and water and salts pass from the interstitial fluid. In this way a finely balanced dynamic equilibrium is set up by water passing from the plasma to the interstitial fluid, from the interstitial fluid to the plasma and from the interstitial fluid to the lymphatics.

Œdema. Œdema, or dropsy, is a disturbance of water balance in which there is an accumulation of fluid in excess in the cells, tissue spaces and serous cavities of the body. Œdema is detected earliest below eyes and behind wrists and is very rare in palms, soles and muscles. Excess of fluid in the peritoneal cavity (ascites) and in the pericardial and pleural cavities (pericardial and pleural effusions) may or may not be associated with generalized œdema, but the mechanisms responsible are essentially the same in principle as those responsible for œdema formation.*

Sodium and chloride disappear from the urine when œdema or dropsy is forming, and œdema is aggravated by the administration of salt. This retention of salt was formerly believed to be due to a primary failure of the kidney to excrete salt. Because the body-fluids necessarily remained isotonic, water retention, leading to œdema, followed. Later the salt retention was regarded as secondary to œdema formation. On that basis the œdema was believed to develop by the various mechanisms considered below, and the salt was diverted from the urine to the interstitial fluid in order to maintain isotonicity. However, it is now believed that in many conditions sodium retention plays an important primary part.

Œdema may be classified into five main types:

1. Inflammatory œdema.   2. Hypoproteinæmic œdema.
3. Congestive œdema.   4. Lymphatic œdema.
5. The œdema of sodium retention.

1. *Inflammatory œdema.* In inflammatory œdema the capillary permeability is increased and the tissue-fluid protein content increased to 1 g or more per 100 ml. The osmotic pressure of the interstitial fluid is thus increased and antagonizes the effect of the plasma proteins. A new equilibrium is set up and water is retained in excess in the interstitial fluid. Examples of this type of œdema are provided by local inflammation of all kinds, such as is found in a boil or a pleural effusion or cellulitis. Inflammatory œdema is also observed in angio-neurotic œdema, and appears to be due to an unusual sensitivity of the capillaries to quantities of histamine to which the capillaries of normal persons are insensitive. The œdema of the acute nephritis was believed to be an example of inflammatory œdema, but sodium retention is now believed to be responsible (see below).

* Such accumulations of fluid in the serous cavities of the body have long been classified by clinicians into transudates and exudates. Transudates do not clot, are low in specific gravity, contain few cells and are not inflammatory in origin. Exudates, however, clot readily, are of high specific gravity and contain cells, and are essentially inflammatory in origin.

2. *Hypoproteinæmic œdema.* In hypoproteinæmic œdema there is a primary reduction in the osmotic pressure of the plasma proteins. This may be due to a variety of causes, such as depletion of plasma albumin due to starvation, protein deficiency, liver diseases, loss of protein-rich fluids in burns or during the drainage of a chronic empyema, as well as massive albuminuria such as occurs in the nephrotic syndrome. This last condition occurs in Ellis Type II nephritis, in amyloid disease and perhaps as a pure nephrotic syndrome in childhood. Clinical conditions closely simulating the nephrotic syndrome may occur in cardiac failure and in diabetic nephropathy. The capillary permeability is not increased and the tissue-fluid protein concentration is within normal limits. The diminished osmotic pressure of the plasma protein results in an adjustment of the dynamic equilibrium so that water is retained in excess in the interstitial fluid.

Hypoalbuminæmia is an important but not the only factor concerned with this form of œdema. In the nephrotic syndrome spontaneous remissions may occur without an increase in the plasma albumin. Attempts to raise plasma protein by concentrated plasma given intravenously do not always induce diuresis. It has been shown that in starvation œdema there is no constant relation between œdema formation and the concentration of plasma protein or the osmotic pressure. Such discrepancies have been attributed to differences in the sodium content of the diet. Reduction of sodium in the diet or the administration of diuretic to diminish the absorption of sodium, did not always reduce œdema, when present, or prevent it from forming. In rats fed on a low protein diet increases in interstitial fluid volumes occur *before* there is a reduction in plasma protein, and an excess of anti-diuretic substances present in the urine.

3. *Congestive œdema.* The increased venous pressure resulting from venous congestion leads to œdema, firstly, because the faulty venous return causes an increased capillary permeability resulting in the passage of protein into the interstitial fluid, and secondly because the hydrostatic pressure at the venous end of the capillary is increased. For many years it was believed that the œdema of congestive cardiac failure was brought about by this mechanism. On this basis fluid should be transferred from the plasma to the interstitial fluid, the circulating blood volume should fall and the hæmatocrit reading rise. However, the

earliest indications of œdema often precede the rise in venous pressure and there is an increase in the circulating blood volume as shown by the diminution in the hæmatocrit reading. It appears that in congestive cardiac failure the diminishing cardiac output is associated with an increased tubular reabsorption for sodium, either relative or absolute. In the earlier stages there is a reduction in glomerular filtration, brought about probably by the fall of blood pressure accompanying cardiac failure, and when this is associated with normal tubular reabsorption there is retention of sodium. The retention of sodium possibly mediated by Verney's osmoceptors and the anti-diuretic hormone of the posterior pituitary results in retention of water with a raised blood volume and a raised venous pressure. This raised venous pressure should tend to increase the cardiac output, a condition impossible in severe heart disease. It appears probable, however, that not only does increased venous pressure play a part in the œdema formation but also reduced glomerular filtration, sodium retention brought about by increased adrenal cortical activity, anti-diuretic hormone and changes in blood pressure.

The role of sodium retention in congestive cardiac failure receives support from the efficacy of low sodium diets, and of modern diuretics in the œdema of this kind. The tubular reabsorption of sodium is decreased, with the consequent loss of sodium and water. The distribution of cardiac œdema, however, appears mainly to be determined by the rise in venous pressure.

4. *Lymphatic œdema.* The blockage of lymphatics by tumour cells, by parasites or by inflammation as in whiteleg after post-partum thrombosis, is well known.

5. *Œdema of sodium retention.* This is observed in its purest form in the sodium retention and with over-dosage with desoxycorticosterone acetate, 9α-fluorohydrocortisone or other mineralo-corticoid. Primary hyperaldosteronism due to an adrenal tumour (Conn's syndrome) is associated with potassium loss, the retention of sodium induced in the early stages does not persist and œdema does not occur. However, secondary hyperaldosteronism is an important factor in the genesis of œdema, and decrease in œdema in the nephrotic syndrome, spontaneous or due to therapy, is often associated with reduction of aldosterone output. In acute nephritis there was formerly

believed to be a generalized increase in capillary permeability leading to the inflammatory œdema. The protein content of œdema fluid in nephritis is, however, not significantly higher than is found in congestive cardiac failure, the earlier high figures being due to contamination with blood plasma. It has also been found that the œdema of acute nephritis was associated with a rise in blood volume and a fall in the hæmatocrit reading. This could not be explained on the basis of an increased capillary permeability, which should result in a fall in the circulating blood volume and an increase in the hæmatocrit, as is observed in angio-neurotic œdema. It appears more probable that in the early stages of acute nephritis the reduced glomerular filtration rate due to the pathological changes in the glomeruli is associated with a relatively normal tubular reabsorption of sodium. The resulting retention of sodium appears to be the cause of the œdema.

Ascites in cirrhosis of the liver was believed to be caused by hypoproteinæmia and to a congestive œdema due to a raised portal pressure.

The hypoproteinæmia was considered to be due to impaired protein synthesis because of liver damage, to be exaggerated by the repeated paracenteses performed in this condition, with subsequent loss of protein in the withdrawn fluid. It is now known, from experiments in which various tributaries to the portal vein and to the hepatic veins have been ligatured, that congestion of the hepatic vein is more important than congestion of the portal vein. On this basis the ascites of cirrhosis of the liver should result from an increase in the interstitial fluid from the liver rather than an increase of the interstitial fluid and lymph from the gut generally. The interstitial fluid in equilibrium with the plasma of the blood vessels of the intestine is similar to interstitial fluid from other parts of the body. It is colourless and of low protein content, but the interstitial fluid and lymph from the liver is of high protein content and yellow in colour. This high protein content of the interstitial fluid in the liver is necessary in order to compensate for the low hydrostatic pressure at the proximal ends of the capillaries supplied by the portal vein. If hepatic interstitial fluid and lymph contained normal amounts of protein this low hydrostatic pressure would be insufficient to balance the unopposed osmotic pressure of the plasma proteins. The ascitic fluid in cirrhosis of the liver is yellow and of high protein content, thus supporting the view that congestion of the hepatic veins is primarily concerned with the development of ascites in this condition. It is interesting to note that portal-caval anastomosis, in which the congestion of the portal vein is relieved, is of greatest value in cirrhosis accompanied by œsophageal varices and hæmorrhage. The operation does not relieve ascites. There is little doubt that the factors concerned in the formation of ascites due to cirrhosis of the liver include a raised hepatic vein pressure and a low plasma albumin, as well as sodium retention, possibly due to the failure of the damaged liver to metabolize or destroy mineralocorticoid hormones.

It is remarkable that the interstitial space in the normal individual is extremely constant in volume, whereas on the classical hypothesis slight changes in colloid osmotic pressure or slight increases in the hydrostatic pressure should lead to progressive alterations in the volume of the interstitial space. Some authorities have therefore proposed that the interstitial space is occupied by a gel which has a swelling pressure and which, like a sponge taking up water, tends to attract aqueous solutions into itself up to a constant limit, so that under most physiological conditions the volume of interstitial fluid does not change much. In conditions leading to œdema formation the gel becomes saturated, the swelling pressure becomes zero, absorption of fluid ceases and further accumulation of water leads to the accumulation of free fluid in the tissues. In the nephrotic syndrome there is no swelling pressure available (the sponge is saturated) to oppose the tendency of fluid to shift to and from the plasma. This makes regulation of the volumes of plasma and of interstitial fluid more precarious than normally.

It has been claimed that in the nephrotic syndrome the increase in colloid osmotic pressure resulting from an increase in the plasma albumin concentration is less than normal. Accordingly, when ultrafiltration is increased, as by an increase in hydrostatic pressure, in the normal subject the resulting hæmoconcentration leads to an increase in the colloid osmotic pressure which opposes further ultrafiltration. In the nephrotic syndrome, however, the colloid osmotic pressure is not increased so much, and therefore ultrafiltration continues unopposed. The result is that a small rise in capillary pressure tends to cause formation of œdema more readily than normal.

The role of aldosterone in œdema. Aldosterone and its metabolites are not easy to determine quantitatively, and since only a very small fraction of the quantity secreted by the adrenal gland is excreted unchanged in the urine, the significance of measurement of aldosterone excretion in urine is difficult to assess. The excretion is increased in some cases of ascites due to cirrhosis of the liver, as well as in pregnancy and in patients receiving low salt diets. It occurs less frequently in the œdema of congestive cardiac failure. There is little doubt that lowering of plasma colloid osmotic pressure plays an important part in the œdema of the nephrotic syndrome. In this condition the hæmatocrit is high, suggesting that there is transudation of fluid from the vascular to the interstitial compartment. This must lead to a diminished blood volume, which influences the "stretch receptors", which then lead to a secondary hyperaldosteronism. In cardiac failure the blood volume is maintained unless energetic diuretic treatment has resulted in peripheral circulatory failure and secondary hyperaldosteronism. A possible sequence of events may be proteinuria leading to loss of plasma volume, increased secretion of aldosterone, causing retention of electrolytes, increase in plasma electrolyte with a secondary increase in anti-diuretic hormone, with consequent retention of water.

The relationship between aldosterone secretion by the adrenal cortex and that of anti-diuretic hormone by the posterior pituitary gland is obscure and may well differ in various clinical syndromes. Aldosterone secretion may be secondarily stimulated by primary retention of water due to antidiuretic hormone, or antidiuretic activity may be secondary to hypertonicity brought about by aldosterone.

The biochemical basis of treatment of œdema and ascites.

Restoration of the protein intake to normal or high levels is the logical form of treatment of œdema of malnutrition or starvation, and in a few cases of massive albuminuria in the nephrotic syndrome in which this treatment does not lead to nitrogen retention due to renal failure. It would seem probable that intravenous administration of concentrated solutions of human albumin or of dextran of suitable molecular weight should increase the colloid osmotic pressure of the plasma and reduce œdema. This is not always the case, either because the infused colloid is rapidly removed from circulation or because there results an increased blood volume leading to cardiac failure rather than a diuresis. If cardiac failure can be avoided some difficult cases may be satisfactorily treated with continued concentrated infusions of plasma albumin. Adrenocorticotrophic hormone (ACTH) or cortisone has been employed empirically to reduce the proteinuria of the nephrotic syndrome.

The limitation of sodium entry into the body either by dietary restriction of salt or by the administration of ion-exchange resins to replace dietary sodium ions by hydrogen or ammonium ions has proved unacceptable with the availability of modern synthetic diuretics.

Mercurial diuretics. The mercurial diuretics combine with the plasma proteins and are not excreted by glomerular filtration; they form complexes with sulphydryl enzymes in the renal cells and are ultimately secreted after metabolic change by tubular secretion. Although they interfere with active absorption throughout the whole nephron, their main effect is on the renal absorption and transport of sodium in the proximal tubule. This results in a larger volume of isotonic fluid than normal passing from the proximal tubule to the loop of Henle, and thence to the distal tubule. Thus, an abnormally large amount of sodium ions is presented to the distal tubule for exchange by hydrogen and potassium ions of sodium ions; chloride is unaffected, and is therefore in higher concentration than sodium in the urine. Mercurial diuretics thus lead to increased excretion of sodium, hydrogen and potassium ions, an excretion of chloride greater than that of sodium with a secondary excessive excretion of water, a mild potassium deficit and a non-respiratory alkalosis due to excessive loss of hydrogen ion. Irrespective of the glomerular filtration rate, about 20 per cent. of the sodium filtered at the glomerulus is excreted in the urine at the height of the diuresis, so that the efficacy of mercurial diuretics is diminished by reduction of the glomerular filtration rate.

Mercurial diuretics are therefore less effective in severe congestive cardiac failure or when there is reduced glomerular filtration rate due to primary renal disease or secondary sodium deficiency, perhaps brought about by previous diuretic therapy.

Carbonic anhydrase inhibitors. The carbonic anhydrase inhibitors, widely used before the introduction of the thiadiazine diuretics, act by depressing the availability of hydrogen ions for exchange with sodium ions in the distal tubule (see p. 6). The depression of hydrogen exchange throughout the whole nephron causes an increased excretion of sodium and bicarbonate ions without effect on the chloride output. The increased sodium load presented to the distal tubule and the reduction of hydrogen-ion

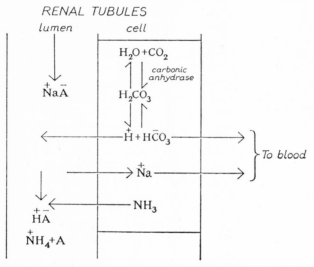

FIG. 5. Transfer of sodium, bicarbonate and hydrogen in renal tubules.

exchange increases potassium exchange, and potassium loss can be marked. This loss of potassium may be exaggerated by secondary hyperaldosteronism. The urine becomes highly alkaline, and a non-respiratory acidosis results from the lowered concentration of bicarbonate and increased concentration of chloride in the plasma. The carbonic anhydrase inhibitors are less effective when the plasma bicarbonate is low, whether due to long-continued treatment or to other causes.

Thiadiazine diuretics. Most modern diuretics are derivatives of benzothiadiazine and have two separate effects. In low dosage they depress the absorption of sodium and chloride ions in the proximal tubule, while in higher dosage they act as carbonic anhydrase inhibitors. Of those two mechanisms the first is the more important and is different from that brought about by the mercurial diuretics, since the effects of the two kinds of diuretics are additive. Of sodium reabsorption by the tubule, 20 per cent is inhibited by the mercurials, 15–20 per cent by the benzothiadiazine derivatives, while 60–65 per cent is unaffected by either drug. These drugs have been particularly effective in the treatment of the œdema of congestive cardiac failure and the nephrotic syndrome, but are less so in that associated with liver disease because of the susceptibility of such patients to potassium deficiency. They are contra-indicated when gout is present, since they can cause an increase in plasma uric acid without influencing the concentration of any other fraction of the non-protein nitrogen.

Aldosterone antagonists. Increased aldosterone secretion by the adrenal may contribute to the abnormal salt and water retention of œdematous patients, although œdema is not produced by excessive aldosterone secretion alone. Aldosterone antagonists, therefore, are not powerful diuretics, but are important potentiators of other diuretics in conditions in which aldosterone is secondarily secreted in excess. Spironolactone resembles aldosterone in structure, and competitively antagonizes its action in the distal part of the nephron, where sodium with accompanying chloride is actively absorbed and exchanged for potassium ions. Such an aldosterone antagonist increases the excretion of sodium and decreases that of potassium. Its action depends on the passage of a relatively large amount of sodium to the distal part of the nephron for its full effect, and this may be achieved by the simultaneous administration of a mercurial or thiazide diuretic.

The administration of many modern diuretics results in a large loss of sodium and a smaller loss of potassium ions. In spite of persisting œdema, excessive diuretic therapy may lead to salt deficiency, with consequent peripheral circulatory failure, pre-renal azotæmia and even coma. The loss of potassium ions in patients on normal diets does not usually lead to potassium deficiency, which, however, may develop after long-continued treatment of patients receiving poor diets. Death may occur unless the deficiency is recognized and prevented.

FURTHER READING

PETERS, J. P. 1944. "Water Exchange." *Physiol. Rev.*, **24**, 491.
—— 1950, "Sodium, Water and Œdema." *J. Mt. Sinai Hosp.*, **17**, 159.
—— 1952, "The Problem of Cardiac Œdema." *Amer. J. Med.*, **12**, 66.
ROBSON, J. M. and STACEY, R. S. 1962. "Recent Advances in Pharmacology," Chapter 7. Diuretics and Electrolyte Balance. p. 214. Churchill, London. (3rd edition).
WILSON, G. M. 1963. The Bradshaw Lecture. "Diuretics." *Brit. med. J.*, **1**, 285.

CHAPTER IV

FLUID BALANCE—SALT DEFICIENCY AND WATER DEFICIENCY

In the normal person the total quantity of body water is maintained fairly constant by a precise adjustment of the total output of fluid and the total intake of fluid. Since under ordinary conditions the loss of fluid by extra-renal routes varies relatively little, the volume of urine is the main variable concerned with balancing fluid output with total fluid intake. Several important points, however, emerge from consideration of the charts of fluid balance shown in Table 5.

TABLE 5—WATER BALANCE FOR AN ADULT AND FOR AN INFANT

Water Balance for Adult of Average Size

Intake		Output		
		Saliva	1,500 ml.	All these fluids are derived from extra-cellular fluid. Compare their volume with 3,500 ml. plasma and 14,000 ml. extracellular fluid.
		Gastric secretion	2,500 ml.	
		Bile	500 ml.	
		Pancreatic juice	700 ml.	
		Intestinal secretion	3,000 ml.	
			8,200 ml.	
As fluid drinks—say	1,450 ml.	Almost all reabsorbed except		normally 100 ml.
As moisture in food	800 ml.	Insensible perspiration		400 ml.
H₂O formed by oxidation of food	350 ml.	Sensible perspiration		200 ml.
		Respiration		400 ml.
		Urine		1,500 ml.
	2,600 ml.			2,600 ml.

Water Balance for Infant of 8 lb.

Intake		Output	
		Alimentary secretions (approx.)	400 ml.
As fluid drinks—say	620 ml.	Almost all reabsorbed normally except	20 ml.
H₂O formed by oxidation of food	80 ml.	Insensible perspiration, sensible perspiration and respiration	180 ml.
		Urine	500 ml.
	700 ml.		700 ml.

1. *The inevitable water loss.* This amounts to approximately 1 litre per day and takes place in the expired air and in the

perspiration. In temperate climates the amount of salt lost in the sensible perspiration, which is hypotonic in comparison with the body-fluids, is quite small, and for all practical purposes the inevitable water loss is one of pure water only. In tropical or other conditions of high environmental temperature this inevitable water loss may be even higher, and may be accompanied by some loss of salt in the sensible perspiration. It is important to realize that this inevitable water loss diminishes with reduction in body-size and amounts to only about 180 ml. per day in the case of an infant.

2. *Water of food.* A normal diet entails the ingestion of about 800 ml. of water as moisture of food. If the patient is existing on a fluid diet or is fasting this amount of fluid will not be available.

3. *Metabolic water.* The metabolic oxidation of the hydrogen of foodstuffs makes available about 350 ml. of water. Even when the calorie value of the food intake is reduced, this source of water will still be available by the oxidation of the hydrogen of the constituents of the tissues.

4. *Reabsorption of gastro-intestinal secretions.* All but about 100 ml. of the 8,000 ml. of gastro-intestinal secretion is normally reabsorbed by the gut.

It is useful to divide fluid loss into three different types:

1. Simple water loss—water deficiency.
2. Loss of electrolyte-containing fluid (e.g., gastro-intestinal secretion)—salt deficiency.
3. Loss of both water and of gastro-intestinal secretion—water and salt deficiency.

The treatment of simple water deficiency is quite different from that of salt deficiency, and it is of the utmost importance that the two conditions should be clearly distinguished, for treatment of the one condition by means appropriate for the other may be lethal. In those conditions in which there is both water and salt deficiency it is imperative to assess roughly the relative contributions from these two factors in these cases.

Water deficiency. A person will become water deficient if he is in coma, if he is unable to swallow because of œsophageal obstruction or because he has no access to water. Not only will he inevitably lose 1 litre of water per day in perspiration and expired air but he will also lose water in the urine, the volume of which is reduced to about 500 ml., the minimum required to eliminate the end products of metabolism. The inevitable water loss is one of pure water, for the small amount of sensible perspiration normally contains very little salt. This water loss may in the first place be regarded as borne by the extracellular fluid. The absence of any corresponding loss of salts must mean that the extracellular fluid becomes concentrated. The excretion of sodium and chloride ions in the minimal quantities of urine secreted alleviate only to a very slight extent this hypertonicity

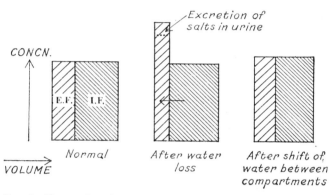

FIG. 6. Changes in volume and concentration of extracellular (E.F.) and intracellular (I.F.) fluid during water loss.

of the extracellular fluid. However, water will pass from the isotonic intracellular fluid into the hypertonic extracellular fluid until the osmotic concentrations inside and outside the cells are equal (Fig. 6). In this way the water loss is shared by both extracellular and intracellular fluids (i.e., by the whole body-water). As will be seen below, this is not the case when the fluids lost contain electrolytes.

Salt deficiency. Primary salt deficiency not only occurs in those conditions in which gastro-intestinal secretions are obviously lost to the body, as by gastric lavage, vomiting, diarrhœa and through fistulæ; it is also an important feature of obstruction of the gut, whether due to pyloric stenosis, acute dilatation of the stomach, intestinal obstruction or paralytic ileus. The obstructed gut continues to secrete the digestive secretions, although reabsorption is completely abolished. Obstruction to the large intestine, however, does not lead to salt deficiency, for the distension of the gut is prevented from extending back to the small intestine by the ilio-cæcal valve. Salt deficiency also occurs during diuretic therapy due to excessive loss of sodium, in Addison's disease and after adrenalectomy, in which the deficiency of adrenal cortical hormones leads to a decreased reabsorption of sodium and chloride ions in the renal tubules.

Gastro-intestinal secretions contain electrolytes in the same osmotic concentration as the extracellular fluid from which they are clearly derived. The extracellular fluid will be reduced in volume, but will in the first place remain isotonic. There can therefore be no compensating shift of water from the cells to the extracellular fluid, so that the loss of gastro-intestinal fluid is borne entirely by the extracellular fluid. Most people who have lost intestinal secretions will have taken fluids and will usually have replaced the electrolyte-containing fluid by one deficient in electrolytes. The extracellular fluid will then become hypotonic, and an osmotic shift of water will now take place from the extracellular fluid to the intracellular fluid (Fig. 7). This is clearly undesirable, for it exaggerates the reduction in extracellular fluid volume. The fall in electrolyte concentration in the plasma results in the absence of sodium and chloride ions in the urine. The continued secretion of the salt-free urine makes the extracellular fluid slightly less hypotonic than it would otherwise be, but results in a persistence of the diminution in volume of the extracellular fluid. It is clear, therefore, that the loss of gastro-intestinal secretion is borne entirely by the extracellular fluid (even though this is replaced by fluid), unless the replacing fluid contains electrolytes and is isotonic with the body fluid. The extracellular fluid rapidly reaches a dangerously low volume, the plasma volume falls, there is peripheral circulatory failure and a prerenal azotæmia occurs early. This is in marked contrast with the state of affairs in simple water loss, which is shared by the entire body-water and in which serious reduction in extracellular fluid and plasma volume occurs very late.

Table 6 illustrates the difference between the effects of water deficiency and those of salt deficiency by contrasting the two conditions in an adult man of, say, 11 stone. Deprived of water, he will lose about 1 litre via his inevitable water loss (i.e., 6 litres of water in about 6 days). His intracellular fluid, normally amounting to about 36 litres, will provide 4 litres of the loss, the remaining 2 litres being sustained by his 14 litres of extracellular fluid. Such a diminution in extracellular fluid volume is not of great moment. On the other hand, if he loses 6 litres of intestinal secretion the intracellular fluid volume will be unchanged and

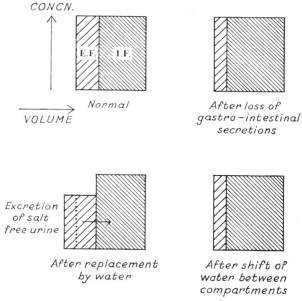

Fig. 7. Changes in volume and concentration of extracellular (E.F.) and intracellular (I.F.) fluid during loss of gastro-intestinal secretions.

the loss of fluid will be borne entirely by the extracellular fluid, the volume of which will fall to only 8 litres. Such a major reduction in extracellular fluid volume will lead to a reduction in plasma volume of such severity that there will be a failure of the peripheral circulation, and a pre-renal azotæmia will develop. The condition is clearly much more serious than is the case in simple water loss.

This difference between water deficiency and salt deficiency, important as it is in the adult, is even more important in children and in infants. Thus a child of 8 lb. who is deprived of water for 48 hours will lose about 360 ml. of water in the expired air and insensible perspiration. The 2¼ litres of intracellular fluid will lose 240 ml. water and the 720 ml. of extracellular fluid only 120 ml. This is well tolerated. On the other hand, the same child may lose 360 ml. of intestinal secretion in a few hours; this loss would be sustained entirely by the 720 ml. of extracellular fluid and would inevitably be fatal.

TABLE 6—DISTRIBUTION OF WATER LOSS IN WATER DEFICIENCY AND IN SALT DEFICIENCY

	Intracellular fluid.		Extracellular fluid.	
	Before.	After.	Before.	After.
Adult 70 kilos.				
Deprived of water for six days: Total H_2O loss = 6,000 ml.	36 l.	32 l.	14 l.	12 l.
Loss of 6 l. of intestinal secretion: Total H_2O loss = 6,000 ml.	36 l.	36 l.	14 l.	8 l.
Infant, 8 lb.				
Deprived of water for six days: Total H_2O loss = 360 ml.	2¼ l.	2 l.	720 ml.	600 ml.
Loss of 360 ml. of intestinal secretion: Total H_2O loss = 360 ml.	2¼ l.	2¼ l.	720 ml.	360 ml.

Loss of highly acid or alkaline intestinal secretions. The alkalosis or acidosis resulting from loss of highly acid gastric juice (as in pyloric stenosis) or of alkaline bile (as with a biliary fistula) are both capable of complete correction by saline therapy alone, and the treatment of this type of disturbance of acid–base balance with acids or alkalis is positively harmful, since it does not correct the fundamental abnormality, which is a loss of salt. The hydrogen and chloride ions lost in acid gastric juice are derived ultimately from the extracellular fluid. The loss of chloride ions leaves an excess of sodium ions in the plasma and extracellular fluid. Similarly, the loss of hydrogen ions in the gastric juice leaves an excess of hydroxyl ions which combine with carbon dioxide, of which there is a sufficiency for forming bicarbonate ions. There results an excess of sodium bicarbonate in the extracellular fluid. Except in the terminal stages, this excess sodium bicarbonate is excreted in the urine. Thus, the net result is that chloride ions are lost in the vomit and sodium ions in the urine. In other words, the body has lost sodium chloride just as if the vomit had not been acid and had contained salt in the same concentration as in the extracellular fluid.

Similarly, if the body loses sodium bicarbonate via a biliary fistula the resulting bicarbonate deficiency in the extracellular fluid is associated with a chloride excess. These chloride ions are excreted in the urine, usually associated with an increased excretion of ammonia. Here sodium is lost in the bile and chloride in the urine, and again the net loss is of sodium chloride. Such disturbances of acid–base balance are correctly treated just as if the gastro-intestinal secretion lost consisted of extracellular fluid.

Treatment of water and electrolyte deficiency. The treatment of patients with water and/or salt deficiency must be based upon the clinical condition of the patient as well as upon the results of hæmatocrit determinations and of analyses of plasma for electrolytes and proteins. The proteins may be estimated very rapidly and conveniently by the specific gravity method. Of the electrolyte determinations, the sodium, potassium and chloride estimations are the more important, but the results of bicarbonate estimation if interpreted correctly (see page 26) are of value. Estimation of the blood urea is also essential to indicate the severity of pre-renal impairment of renal function.

Treatment must have three objectives: (1) the replacement of existing deficits; (2) the supply of the basic requirements of electrolytes and water; (3) replacement of continuing losses.

Water deficiency. Simple water deficiency will in the first place be indicated by the presence of a clinical condition known to lead to water deficiency, and by thirst. Laboratory confirmation will be obtained by raised values for serum electrolytes, plasma proteins and, in the absence of bleeding or a hæmatological disorder, by a raised hæmatocrit.

Patients with simple water deficiency require fluid drinks by mouth. Fluid should be given until the daily output of urine is at least 1 litre. Intravenous therapy should be avoided unless there is inability to swallow, when 5 per cent glucose in water should be given. The glucose will provide calories, usually much needed by such patients, and will be metabolized within a few hours. Saline must never be given to these patients, for the extra salt will increase the hypertonicity of the extracellular fluid, and its elimination in the urine will require extra water, so that the dehydration becomes exaggerated.

Repeated hæmatocrit or serum protein investigations and perhaps an occasional estimation of plasma electrolytes are all that are required to control fluid therapy, but when the patient has been adequately hydrated a determination of the electrolytes might provide a safeguard against water intoxication due to overhydration. Water intoxication can occur especially readily during oliguria and will be shown clinically by anæmia, lethargy, bradycardia and finally by convulsions and delirium. It is one of the few conditions requiring treatment by hypertonic saline.

Salt deficiency. In primary salt deficiency the patient is ill and lethargic and not usually thirsty. The sodium and chloride concentrations in the plasma are usually low, while the plasma protein and hæmatocrit values are high.

It is important to calculate roughly the volume of the extracellular fluid of the patient requiring treatment, and if it is appreciated that a patient whose extracellular fluid volume is depleted by much more than a third would probably be dead, a good many fatalities due to over-treatment with enormous quantities of saline would be avoided. Thus a 70-kilo man would have an extracellular fluid volume of 14 litres, and in the absence of a continuation of the primary salt loss—e.g., by a persistence of the intestinal obstruction or of vomiting—would be unlikely to require more than 5–6 litres of normal saline. If he does require more than this, it is essential that his treatment should be controlled by frequent estimations of electrolytes and of plasma protein or hæmatocrit.

If possible, 0·9 per cent saline flavoured with lemon and glucose should be given by mouth but usually the cause of the condition is such that this is impracticable. Then 0·9 per cent saline with glucose must be given intravenously. Ideally, the total loss of intestinal secretion should be measured and precisely the same amount of glucose–saline given, but when the patient has lost unknown quantities of gastro-intestinal secretion either before admission to hospital or by retention in a dilated gut this may not be possible.

If there has been no recent operation it is sometimes possible to control saline therapy by qualitative examination of the urine for chlorides. Intravenous therapy with 0·9 per cent saline with glucose should be continued until chlorides appear in appreciable quantities in the urine, i.e., until there is a flocculent precipitate of silver chloride in the usual qualitative test for urinary chlorides; then if the salt loss is no longer continuing, the 0·9 per cent saline with glucose should be replaced by 0·2 per cent saline with glucose. Some workers prefer to estimate the chlorides in the urine, but the author is of the opinion that the qualitative test is sufficient for most purposes. However, urinary chloride determinations do not always reflect the balance of chloride and sodium in the body, because after surgical operations some patients with hyperchloræmia may excrete very little chloride and others with hypochloræmia may excrete much chloride. The former situation may also occur after head injuries. For this reason, it is best to control intravenous saline therapy by repeated analyses of plasma electrolytes wherever possible.

At least the plasma sodium should be estimated and preferably the plasma chloride and bicarbonate as well, especially when the gastro-intestinal secretions lost are highly acid or alkaline, as in pyloric stenosis and biliary fistula respectively (see below). Treatment should be continued until the electrolyte content of the plasma has been restored to normal. Then, if intestinal secretions continue to be lost they should be measured accurately and replaced by the administration of an equal quantity of 0·9 per cent saline with glucose. Once a salt loss has been corrected or is being corrected in this way, the total fluid intake should be carefully balanced so that sufficient fluid is taken to replace not only the urine loss but also the inevitable water loss via the insensible perspiration and expired air.

It might be thought possible to calculate from the results of the analyses of the plasma electrolytes the amount of 0·9 per cent sodium chloride required to restore isotonicity, but because of the elimination of sodium chloride free urine and the shift of water into the cells, the alteration in volume of the extracellular fluid is always greater than is indicated by the actual alteration in the composition of the plasma. Also, a continued loss of potassium, which may become significant if the loss of gastro-intestinal secretion is prolonged, may result in the intracellular fluid contributing to an unknown extent to the fluid loss. For these reasons the treatment of salt deficiency with intravenous saline is entirely empirical and must be controlled by frequent estimations of electrolytes and guided by clinical assessment of the patient's condition. Nevertheless, a rough assessment of probable requirements from the electrolyte analyses and from the roughly calculated extracellular fluid is essential.

An attempt to calculate saline requirements from the sodium deficiency and the extracellular fluid volume will indicate quantities too low, and calculations by some workers based on the sodium deficiency and the total body water invariably suggest amounts which are too great. A satisfactory but quite empirical indication of likely requirements can be obtained from the chloride deficiency and the extracellular fluid volume in the following way:

If plasma $\overline{Cl}$ = 80 mEq/l., extracellular fluid is only 80/100 of isotonicity (the mean normal plasma $\overline{Cl}$ = 100).

Deficiency = $\dfrac{100 - 80}{100}$ = 20/100 = 1/5.

If extracellular fluid volume (estimated as $\frac{1}{5}$ body weight) = 12 litres the required amount of normal saline = $\frac{1}{5} \times 12$ = 2·4 litres (= approx. 4 pints).

In fact, this provides about 50 per cent more chloride than would be indicated by the chloride analyses and extracellular fluid volume, because isotonic saline contains 150 mEq/l. while normal plasma contains only 100 mEq/l.

Mixed water and salt deficiency. Unless there has been loss of gastro-intestinal secretions, water loss is usually the most marked feature after major surgical operations. Usually there has been restriction of fluid intake leading to the simple water deficiency, but sweating is often excessive, causing a mild degree of salt deficiency, which may be exaggerated by slight post-operative vomiting. This mixed salt and water deficiency is best corrected by the administration of 0·2 per cent saline with glucose, which should be administered by mouth if possible or intravenously if necessary.

The priority of correction of electrolyte and water loss over early surgical intervention differs greatly from case to case. Thus, a patient with pyloric stenosis may come under the surgeon's care only after many days of vomiting. Unless the salt and water balance has been carefully controlled previously, which is unlikely, the patient will be grossly sodium deficient. The extracellular fluid volume and the plasma volume will already be seriously lowered. Immediate surgical intervention would be extremely hazardous owing to the possibility of further lowering of plasma volume by hæmorrhage and shock. Intravenous treatment with 0·9 per cent saline with glucose will restore the extracellular fluid volume to normal, and greatly reduce the surgical risk. Unfortunately, the matter is not always so simple as this.

Table 7 summarizes the suggested lines of treatment for water deficiency, salt deficiency and mixed salt and water deficiency.

TABLE 7—GUIDE TO TREATMENT OF DEHYDRATION

1. Simple H_2O loss	Simple operative procedures accompanied by sweating and restriction of fluid intake Inability to swallow	Fluid by mouth, or if necessary 5% glucose in H_2O intravenously	Hæmatocrit, serum protein and electrolyte analyses until high values have returned to normal. Measurement of urinary output till daily output adequate (at least 1 litre)
2. Mixed H_2O and sodium loss	Operative procedures accompanied by restriction of fluid intake and sweating in theatre Slight post-operative vomiting	Glucose lemonade in 0·2% saline by mouth. If such route impracticable 5% glucose in 0·2% saline intravenously	Measure urinary output. Examine urine qualitatively for chlorides till chloride present in urinary volume adequate. Determine serum electrolytes etc. as below
3. Primary sodium deficiency	Severe vomiting Gastric lavage Pyloric stenosis Acute dilatation of stomach Intestinal obstruction Paralytic ileus Severe diarrhœa Addison's disease Drainage of digestive juices through fistulæ "Salt-free diets" Diuretics influencing sodium excretion	Glucose lemonade in 0·9% saline by mouth if possible. If not, 5% glucose in 0·9% saline intravenously	Determine serum electrolytes until low values have risen to normal and hæmatocrit and serum protein analyses have fallen to normal

Such patients are frequently in poor condition not only on account of the sodium deficiency but also on account of what amounts to starvation. While calories can be provided by the glucose administered with saline, the protein loss is less easy to correct, although protein hydrolysates may prove useful. If the period of pre-operative correction of salt and water balance is made too long the protein starvation may become a major factor. It is therefore sometimes necessary to operate immediately and before salt and water deficiency has been completely corrected. This is especially the case in small intestine obstruction. 0·9 per cent saline with glucose should be given intravenously during the pre-operative period as rapidly as possible without overloading the patient's circulation. This intravenous therapy, which should be continued during and after operation until the salt loss is corrected, may need to be supplemented with whole blood or plasma to correct hæmorrhage or shock. Treatment of shock or of hæmorrhage by plasma or blood transfusion must always take precedence over the treatment of dehydration.

In some circumstances intestinal secretions may be lost via routes which enable the amounts concerned to be accurately measured. This is the case with drainage of a bile duct or when gastric suction is used to keep the stomach empty. The volumes of fluid lost each day should be accurately measured, and exactly this amount of normal saline should be given each day by mouth if this is possible, but intravenously if this is not the case, as with gastric lavage. The rest of the daily fluid requirements, computed according to the balance sheet of Table 5, should be made up with 0·2 per cent saline with glucose.

Salt and water balance in uræmia and oliguria. At no time is an accurate knowledge of salt and water balance more important than in oliguria, whether due to hæmolysis following a mismatched transfusion or following some operation in the region of the pelvis. Here it is essential to remember that all the time there is no vomiting there is no loss of salt, and that although no urine is produced, there is an inevitable water loss. In temperate climates the water intake should therefore be limited to 1 litre per day plus whatever volume of urine is produced, and this is best done by giving a high-calorie, low-protein intake as discussed on pages 15 and 16.

Potassium and potassium deficiency. Recent work has shown that in conditions associated with salt deficiency, potassium ions in fact pass out from the intracellular fluid to the extracellular fluid, tending to increase the concentration of potassium in this fluid compartment. There is therefore an increase in plasma potassium and an increased excretion of the ion in the urine. This transfer of potassium from cells to extracellular fluid may in part lessen the untoward osmotic flow of water in the opposite direction, but to what extent it does so is not known, for in many conditions in which potassium ions pass out of the cells sodium and hydrogen ions pass in. It seems likely that this transfer of potassium from the cells may diminish to some extent the extracellular dehydration which would otherwise occur.

Potassium deficiency is rare and much less common than sodium deficiency, not only because the total body content of potassium is greater than that of sodium but also the gastro-intestinal secretions as a whole contain much less potassium than sodium. Apart from the very rare cases with an aldosterone-secreting adrenal tumour, potassium deficiency of sufficient severity for clinical significance occurs only after long-continued loss of gastro-intestinal secretions, and is then commonly manifest only after treatment has restored the sodium content of the body to normal. This is because the kidneys continue to excrete potassium in spite of a deficiency, unlike the disappearance of sodium from the urine in most cases of sodium deficiency.

No satisfactory laboratory test has so far been devised for the infallible recognition of potassium depletion. Analysis of the serum is of no value, for the serum potassium may be normal or even slightly raised at a time when the body depots are greatly depleted. The determination of potassium content of the red cells does not indicate the potassium content of the cells of the body in general. The urinary output of potassium after administration of potassium salts has been used, but insufficient experience is available for this to be a routine investigation.

Potassium depletion must therefore be diagnosed essentially on clinical grounds. The presence of muscular weakness or paralysis, complete or partial atony of the bowel associated with an abnormality of the electrocardiogram should make one suspicious of the condition, which, however, should never be diagnosed in the absence of a clinical condition known to result in a deficiency of this ion. Thus potassium deficiency should not be thought of unless there has been loss of gastro-intestinal secretion for at least a week, or unless a patient has been maintained for this time on saline–glucose infusions or who otherwise has been living on a diet low in potassium especially during diuretic therapy (see pp. 44–47).

The condition is best treated by the administration of fruit juices or potassium salts by mouth. Intravenous administration of potassium salts can be dangerous, and should be undertaken only if full laboratory facilities are available for repeated measurements of serum potassium and if there is a plentiful flow of urine. Only in these circumstances may 2 g potassium chloride, i.e., about 27 mEq K, be added to each pint (or 500 ml.) of saline–

glucose. The serum potassium should be measured after each pint has been given, and if the serum potassium increases and approaches normal levels the addition of potassium should be stopped immediately. Genuine cases of potassium deficiency may require several days of such treatment before the serum potassium begins to rise.

Magnesium deficiency. As the features of potassium deficiency were revealed when sodium deficiency became correctly appraised and treated, so magnesium deficiency has been observed in subjects maintained on intravenous therapy for long-continued loss of gastro-intestinal secretions or for a serious intestinal absorption deficit. Deficiency of magnesium is probably rare because renal conservation is good, but should be suspected if tetany with psychiatric symptoms develops during long-continued loss of gastro-intestinal secretion. Estimation of serum magnesium is not easy, and is probably best referred to a laboratory specifically interested in this subject.

Electrolyte deficiencies in infancy and childhood. In childhood treatment of salt and water deficiency may be complicated, especially in early infancy, by the failure of the immature kidney to excrete sodium and chloride ions in high concentrations. It is therefore even more important than in the adult that the volumes of the fluid compartments, especially of the extracellular fluid, should be roughly calculated in every case. Otherwise absurd and dangerous quantities of fluid may be given. It is better to administer saline not stronger than half normal, and even this solution should be given only when gastro-intestinal secretions have been lost. In all other circumstances 5 per cent glucose in water or 0·18% saline should be given. The normal maintenance requirements in an infant may be roughly computed on the basis of $2\frac{1}{2}$ oz. per pound body weight (140 ml./kg body weight) per 24 hours. This will need to be exceeded if there has been excess loss of fluid or if that loss of fluid is continuing.

In the past, gastro-enteritis in infancy has probably been the most frequent cause of potassium deficiency, and although hitherto many pædiatricians have used an isotonic intravenous fluid containing a considerable proportion of potassium, the condition is better treated by adding ampoules containing 1 g of potassium chloride to the intravenous fluid, taking great care that urine

SALT DEFICIENCY

output is maintained and that serum potassium is measured before and after the infusion.

Electrolyte patterns in other conditions. Fig. 8 summarizes the alterations in electrolyte patterns of the plasma produced in various conditions.

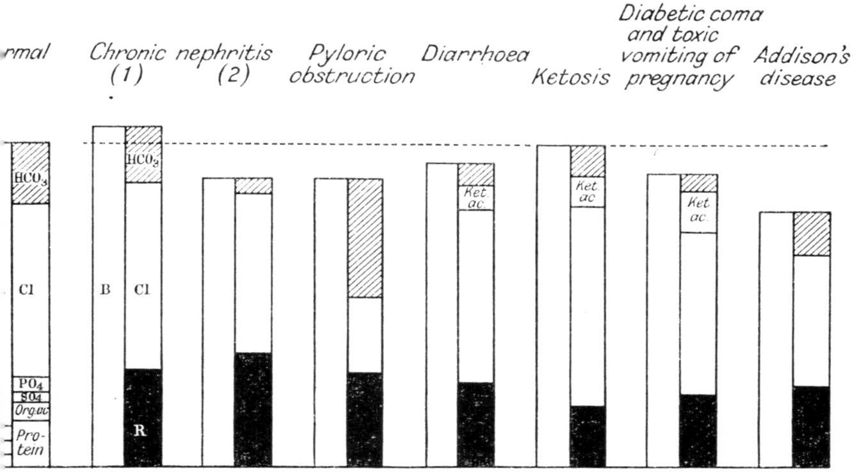

FIG. 8. Alterations of electrolyte structure of plasma produced by various conditions of disease (after Gamble).

A basic list of intravenous fluids

5·4% glucose	approx. 300 milli-osmols/l.
Isotonic saline (0·9 g NaCl%)	150 mEq.Na$^+$/l.
	150 mEq.$\overline{\text{Cl}}$/l.
0·18% NaCl + 4·8% glucose	30 mEq.Na/l. approx. 300
	30 mEq.Cl/l. m-osmols/l.
Isotonic sodium bicarbonate solution	150 mEq.Na$^+$/l.
(1·25% NaHCO$_3$)	150 mEq.H$\overline{\text{C}}$O$_3$/l.
10 ml. ampoules of 10% KCl (for addition to 0·5 l. bottles of other solutions)	13·4 mEq.K$^+$ per ampoule
	13·4 mEq.$\overline{\text{Cl}}$ per ampoule

Further Reading

Black, D. A. K. 1964. "Essentials of Fluid Balance." Blackwell, Oxford (3rd edition).

Taylor, W. H. 1965. "Fluid Therapy and Disorders of Electrolyte Balance." Blackwell, Oxford.

CHAPTER V

LIVER FUNCTION AND DISORDERS OF THE LIVER

Jaundice. Jaundice frequently, but not always, accompanies disturbances of the liver and is due to accumulation of bilirubin or conjugated bilirubin in the plasma and interstitial fluid. The mechanism of production of jaundice can be understood only if the modes of formation and disposal of the bile pigments are known.

The mode of formation and subsequent fate of bilirubin is shown in Fig. 9. Bilirubin is manufactured in the cells of the reticulo-endothelial system from the hæmoglobin of red cells at the end of their life span and is conveyed in the bloodstream to the liver, where it is conjugated mainly with glucuronic acid and probably also with sulphate for excretion in the bile.

The conjugation of bilirubin and its transfer is complicated. The pigment arrives at the surface of the liver cell as a bilirubin–albumin complex; it is not known whether it passes into the cell while still bound to albumin. Conjugation occurs in or near the microsomes, where it reacts with uridinediphosphoglucuronic acid formed in the cytoplasm from uridinediphosphoglucose synthesized in the nucleus. The bilirubin glucuronide then requires transfer from the microsome to the bile. Conjugation with sulphate is probably equally complicated.

In the intestine the conjugated bilirubin is reduced to a mixture of colourless chromogens conveniently referred to as "fæcal urobilinogen". These chromogens are readily dehydrogenated to give orange-red pigments, the so-called fæcal urobilin.* These pigments have very similar chemical properties, as do their respective precursors, and are difficult to separate. "Fæcal urobilinogen" is reabsorbed in considerable amounts and passes in the blood-stream to the liver, where it is said normally to be either metabolized or re-excreted in the bile. In liver disease the

* The terminology of these pigments has been very confusing, and the term " stercobilin " is better used for the only one of the pigments which can most readily be crystallized from fæcal extracts.

reabsorbed urobilinogen is diverted to the kidneys for excretion in the urine.

It is also evident from Fig. 9 that normal blood plasma contains bilirubin, in transport from the reticulo-endothelial system to the liver, which normally excretes it so rapidly that the concentration in the blood is less than 0·7 mg/100 nl. Bilirubin can accumulate in the plasma in slightly increased amounts without

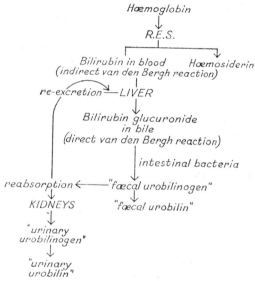

FIG. 9.—Bile pigment metabolism.

the general pigmentation becoming recognizable clinically; the condition is then known as latent jaundice. When the plasma bilirubin reaches concentrations of about 2 mg/100 ml. the general pigmentation of the body tissues can usually be detected by the skilled observer; with concentrations above 3 or 4 mg/100 ml. jaundice is obvious. In severe jaundice the plasma bilirubin frequently reaches levels of 15–20 mg/100 ml., and may attain a concentration as high as 50 mg/100 ml.

Classification of jaundice. Jaundice is now best classified into post-hepatic (obstructive), hepatic and pre-hepatic jaundice, according to whether the causal abnormality is obstruction of the extra-hepatic bile ducts, in the liver, or due to the liver being presented with more bilirubin than can be conjugated.

Post-hepatic (obstructive) jaundice. This is due to obstruction from gall stones, tumour or scarring from infection. The resulting distension of the biliary passages above the site of obstruction was formerly believed to cause rupture of the biliary canaliculi within the liver with regurgitation of bile in the venous sinusoids and passage of conjugated bilirubin into the blood stream. This form of jaundice has therefore been known as mechanical regurgitation jaundice. It is now considered probable that there is an abnormality of transport of conjugated bilirubin from the site of conjugation to the biliary canaliculi; conjugated bilirubin, therefore, passes into the blood stream.

Hepatic jaundice. In this form of jaundice there is localized or more generalized necrosis or death of the liver cells; changes also occur in the surviving cells, which may become swollen. New liver cells arise by division of those cells which survive the initial damage. Until recently hepatic jaundice has been known as parenchymatous regurgitation jaundice and considered to be due to intra-hepatic obstruction (caused by the altered architecture of the liver lobule) leading to regurgitation of conjugated bilirubin into the blood as with post-hepatic (obstructive) jaundice. Investigation of certain rare forms of jaundice (see below) in which plasma contains conjugated bilirubin, together with substantial amounts of unconjugated bilirubin, suggests that accumulation of conjugated bilirubin in the blood plasma in these types of jaundice and in hepatic jaundice is due to impairment of transport of conjugated bilirubin into the bile; the bilirubin, therefore, passes into the blood stream.

Hepatic jaundice may be subdivided into: (*a*) diffuse hepatocellular injury, as in acute and sub-acute hepatitis, acute necrosis; (*b*) due to drugs and toxins; (*c*) intra-hepatic cholestasis which shows many of the features similar to extra-hepatic bile duct obstruction, despite the preservation of the normal lobular architecture.

Pre-hepatic jaundice. This includes hæmolytic jaundice, in which excessive bilirubin is formed from hæmoglobin by excessive intra-vascular or extra-vascular hæmolysis of red cells. In non-hæmolytic pre-hepatic jaundice there is excess conjugated bilirubin in the plasma, but there is no evidence of excessive hæmolysis. These two types of pre-hepatic jaundice have previously been called hæmolytic and non-hæmolytic retention jaundice.

Hæmolytic pre-hepatic jaundice. In hæmolytic jaundice there is an excessive rate of breakdown of red cells and of their hæmoglobin so that bilirubin passes into the bloodstream at a greater rate than that at which it may be removed by the liver. Some workers consider that the normal liver can always remove bilirubin from the blood at a rate sufficient to prevent hyperbilirubinæmia even if breakdown of hæmoglobin is six times the normal and that there must also be impaired biliary secretion for hæmolytic jaundice to occur.

Non-hæmolytic pre-hepatic jaundice. Jaundice can sometimes occur as a hereditary abnormality (Gilbert's disease) in which the plasma contains excess bilirubin, giving a negative direct van den Bergh reaction (see below), and yet there is no evidence of excessive breakdown of red cells. There is failure of the liver to remove bilirubin from the bloodstream because of a congenital deficiency of an enzyme concerned with conjugation (see p. 169).

Other rare varieties of jaundice. Sometimes the plasma of subjects with familial non-hæmolytic pre-hepatic jaundice contains both conjugated and non-conjugated bilirubin. In these subjects there is not only an impairment in the excretion of bilirubin but also of bromsulphthalein. There appears to be an inability to transfer conjugated bilirubin into the bile. The "chronic idiopathic jaundice" of Dubin-Johnson is characterized by the presence of conjugated bilirubin in the plasma and of granules of melanin derivatives in the liver. In an even rarer form of jaundice excessive bilirubin is formed from some source other than the red cells at the end of their life span.

Van den Bergh reaction. When fluids containing bilirubin are treated with alcohol and the van den Bergh reagent (diazotized sulphanilic acid), a red pigment, azobilirubin, is formed. The intensity of the colour provides a measure of the concentration of bilirubin. Van den Bergh found that some sera would react with the diazo reagent even without the addition of alcohol, while other sera, mainly from normal people and from patients with hæmolytic jaundice, required the addition of alcohol before azobilirubin could be formed. Sera which formed azobilirubin in the absence of alcohol were said to give a positive direct reaction, while those forming azobilirubin only in the presence of alcohol were said to give negative direct reactions. Those sera giving positive direct reactions contain conjugated bilirubin, those giving the negative reaction contain only unconjugated bilirubin. The value of the direct reaction lies in that if it is completely negative in the presence of hyperbilirubinæmia one may diagnose hæmolytic or non-hæmolytic pre-hepatic jaundice with certainty. This is because the jaundice is due to the accumulation of unconjugated bilirubin in the blood. A direct positive reaction is of no value in diagnosis, for the circulating bilirubin is partly or wholly in the conjugated form, and this occurs in obstructive jaundice, hepatic jaundice and only rarely in hæmolytic jaundice.

Pigment excretion in jaundice. Conjugated bilirubin is soluble in water and readily excreted into the urine; bilirubin itself is insoluble in water and not excreted by the kidneys, although like the conjugated form, it is held in solution in the blood plasma by linkage with the proteins.

In hæmolytic pre-hepatic jaundice, which may be accompanied by an anæmia if the rate of erythropoiesis lags behind the rate of red-cell destruction, raised amounts of bile pigments are excreted by the liver, and the stools are either normal in colour or more deeply pigmented than normal. Determination of "fæcal urobilinogen" shows an increased excretion relative to the circulating hæmoglobin. The urine does not contain bilirubin, although urobilinogen and urobilin may be present in increased amounts.

In obstructive jaundice less pigment than normal reaches the gut and the fæces become less pigmented (clay-coloured). If the obstruction is due to a malignant tumour much less pigment reaches the gut than if the obstruction is intra-hepatic or due to a stone. However, the fæces may be coloured brown by altered blood, even when little or no bile pigment is present. Bilirubin is always present in the urine, but urobilinogen and urobilin are usually absent unless there is infection of the biliary tract or the obstruction has been present so long that secondary liver damage has occurred.

In hepatic jaundice the stools are also pale or clay-coloured. Bilirubin is found in the urine, and in the early stages may be present even before there is an increase in the plasma bilirubin;

on the other hand, during recovery bilirubinuria may disappear before the jaundice has completely gone. The damaged liver fails to re-excrete into the bile the urobilinogen absorbed from the gut, and urobilinogen and urobilin are present in the urine, except when jaundice is severe, when they are found only during the early stages of the condition and during recovery. When jaundice is severe the gut contains little bile pigment and there is no urobilinogen absorbed for the damaged liver to divert to the kidneys. Table 8 summarizes the interpretation of the results of these tests. Early tests for these pigments in the urine are therefore important in assisting diagnosis, especially before the

TABLE 8—BILE PIGMENTS IN URINE AND FÆCES

	Urine.		Fæces.
	Bilirubin (conjugated).	Urobilinogen and/or urobilin.	Urobilinogen.
Normal	0	Trace	++
Pre-hepatic jaundice:			
hæmolytic	0	Trace or +	++++
non-hæmolytic	0	0	0→+
Obstructive jaundice	+	0	0→+
Hepatic jaundice:			
mild	+	+	0→+
severe	+	0	0
Cholestatic	+	0	0
Obstructive jaundice and infection	+	+	0→+

jaundice becomes severe, otherwise more complicated investigations may be required to distinguish between post-hepatic and hepatic jaundice. However, urobilinogen excretion in the urine increases with increasing pH and although valuable as a qualitative test cannot be put on a quantitative basis.

Tests of liver function. Tests of liver function may help in the differential diagnosis between the three forms of jaundice and between jaundice due to tumour, stone or infection of the biliary tract. They may assist in the assessment of liver function before surgical portal-caval anastomosis for gastro-intestinal hæmorrhage in cirrhosis.

The choice of liver function test is difficult because of the numerous and varied functions performed by the liver cells, by different effects disease may exert on these various functions and more important because liver damage is usually followed immediately by liver regeneration. Liver function must depend upon the balance between damage and regeneration, and since damage to liver cells may continue and liver regeneration persist even after the original factor responsible for the liver damage may no longer operate, the results of liver function tests often change considerably during the course of disease. No one function test can be regarded as more sensitive than another which may assess some completely different function and several tests must be carried out, and these can only investigate and assess a few aspects of the numerous functions of the liver at a particular time.

Liver function tests may usefully be classified into: (i) tests of excretory function; (ii) true tests of liver function; and (iii) empirical tests. In addition, there are special tests helpful in the differential diagnosis of hepatic coma (see p. 79). Measurements of certain enzymes in the blood may be of great value, although the precise concentration may depend upon a variety of factors, including excretion, liberation of enzymes from cells due to increased permeability as well as increased enzyme synthesis, perhaps associated with regeneration.

Tests of excretory function. Numerous tests have been devised to assess the excretory function of the liver. Substances used have included bilirubin, stercobilin, bromsulphthalein, etc. The bromsulphthalein test is useful, especially when jaundice is absent. The material is given intravenously, and the amount which has been retained in the circulating blood is measured after a certain time.

The liver is concerned with the secretion of alkaline phosphatase into the bile, and in both hepatic and obstructive jaundice the plasma alkaline phosphatase is above normal. In pre-hepatic jaundice the serum phosphatase is usually within normal limits. A high percentage of cases of obstructive jaundice show phosphatase levels of more than 30 units/100 ml., and a high percentage of cases of hepatic jaundice show phosphatase levels of 13–30 units/100 ml. The reason for this difference is not understood, although lowered synthesis of the enzyme may play some part.

The plasma concentrations of many enzymes, especially glutamic pyruvate transaminase (SGPT) as well as glutamic oxaloacetic transaminase (SGOT), are considerably increased in conditions associated with the necrosis of liver cells and to a lesser extent in obstructive jaundice. These investigations, however, are not specific, because increases of both enzymes, especially of the SGOT, occur after cardiac infarction (see p. 173). Most laboratories prefer the estimation of SGPT because higher levels are attained; others prefer lactic dehydrogenase, which remains elevated for a longer time and may be used to assess prognosis. Other enzymes, such as aldolase, lactic, isocitric and glucose 6-phosphate dehydrogenases, are also increased in the plasma in liver damage.

Jaundice is usually associated with increased plasma concentrations of cholesterol, which is normally excreted in the bile, but estimations of plasma cholesterol are of little value in the differential diagnosis of jaundice, although high values are found in the intrahepatic cholestasis of drug jaundice.

True tests of liver function. The liver plays an important role in glycogenesis, and the conversion of glucose into glycogen may be profoundly altered in liver disease. Thus, in advanced hepatic failure there is hypoglycæmia, while in less severe degrees of liver insufficiency the glucose tolerance is typical of diabetes. However, the tolerance to other carbohydrates, such as lævulose and galactose, is also radically decreased, but extra-hepatic factors may profoundly alter carbohydrate tolerance so that tests of this function are now rarely used in the diagnosis of hepatic disease.

In acute necrosis of the liver there is failure of deamination of amino acids by the liver, leading to a decreased blood urea and an increased excretion of amino acids in urine. This last occasionally results in crystallization of the less soluble amino acids, leucine and tyrosine, from the urine when there is very severe liver damage. The identification and rough estimation of the amino acids by paper chromatography may permit recognition of failure of deamination in less severe degrees of liver failure.

The liver cells play an important role in protein synthesis, and the concentrations of albumin, cholinesterase and prothrombin in plasma are diminished in the presence of liver disease; their estimation may be of value in the differential diagnosis of jaundice. Plasma albumin is often decreased in hepatitis or other forms of liver damage even of only moderate severity, and may be slightly decreased in protracted obstructive jaundice. The cholinesterase concentration is also diminished in liver disease but not in obstructive jaundice, and has been used as a diagnostic test. Hypoprothrombinæmia associated with liver damage may, of course, be distinguished from hypoprothrombinæmia associated with obstructive jaundice. In this last condition the "prothrombin time" (which varies inversely as the prothrombin concentration) rapidly returns to normal after parenteral vitamin K therapy, while in liver disease the prothrombin time remains prolonged despite adequate parenteral administration of vitamin K.

Empirical tests of hepatic function. There is sometimes a significant increase in the total globulin of the plasma, but more often there is simply an increase in the γ-globulin. Zone electrophoresis, although not always providing quantitative results in agreement with classical methods, can give valuable infomation concerning the distribution of the main fractions of the plasma proteins. When the qualitative partition of these proteins becomes abnormal certain empirical precipitation and flocculation reactions become positive. These may give as valuable information as the more detailed investigation by electrophoresis. The thymol turbidity, the ammonium sulphate and the zinc sulphate tests are the most useful and although not positive in all cases of liver damage, are seldom positive in its absence. However, in multiple myelomatosis the zinc sulphate test is usually more strongly positive than the thymol turbidity test (see p. 87).

Differential diagnosis of hepatic and obstructive jaundice. When bilirubinuria is present, urobilin or urobilinogen must be looked for in the urine during the early stages of jaundice or there may be difficulty in differentiating obstructive jaundice from acute hepatitis. A combination of tests, usually the phosphatase and thymol turbidity or other flocculation test, will often enable a diagnosis to be correctly reached in the majority of patients. In a few cases other tests may be necessary. The differential estimation of the serum proteins by electrophoresis is probably the most useful in these circumstances, acute hepa-

titis usually being accompanied by an actual increase in the
γ-globulin as well as by a fall in the albumin concentration. In
the acute stages of hepatitis and in liver damage generally, the
serum enzymes may be considerably increased; they are also
increased to a lesser extent in obstructive jaundice. Sometimes,
in spite of all possible combinations of tests, an equivocal result
is obtained, and then biopsy or laparotomy may be necessary.
Determination of fæcal urobilinogen may assist in the diagnosis
of obstructive jaundice due to carcinoma. In this case obstruction is almost always complete and the "fæcal urobilinogen" is
usually less than 5 mg/day. In contrast, obstruction due to
calculus is seldom complete, the stone tending to act as a valve
leading to a fæcal pigment excretion of 10 or more mg/day.
Cases of acute hepatitis at this stage usually excrete 5–10 mg/day.

Liver Function

Hæmolytic pre-hepatic jaundice. The laboratory findings in
uncomplicated hæmolytic jaundice are characteristic. The
blood contains excess unconjugated bilirubin, as shown by the
negative direct van den Bergh reaction. The urine contains no
bilirubin nor any excess of urobilinogen or urobilin, the stools
are very dark and contain much urobilin and urobilinogen.
Quantitative determination of the total fæcal urobilinogen shows
that there is more than 20 mg/100 g of circulating hæmoglobin.
There is hæmatological evidence of increased hæmolysis, and
anæmia with reticulocytosis is usually present. There may be
spherocytosis, increased fragility or other abnormality of the
erythrocytes (see pp. 195–198).

Sometimes in hæmolytic jaundice the serum contains both
conjugated bilirubin and unconjugated bilirubin, as shown by a
positive direct van den Bergh reaction. Urobilinogen and urobilin and sometimes traces of bilirubin are present in the urine.
The other findings in the blood and stools are the same as when
there is no liver damage. Other liver function tests do not
usually indicate abnormality.

In hæmolytic jaundice pigment stones may form in the biliary
passages owing to large amounts of pigment excreted. The
diagnosis may then be extremely difficult, for obstruction
may reduce the amount of stool pigment. The condition should
be suspected in any young person with symptoms of biliary

calculi or whenever pure pigment stones are found at operation.

Non-hæmolytic pre-hepatic jaundice. In non-hæmolytic pre-hepatic jaundice the laboratory findings are similar to those of hæmolytic jaundice without liver damage, but there is no hæmatological evidence of excessive hæmolysis.

Acute hepatitis. Acute hepatitis is a disease of varying severity and duration, and Fig. 10 represents what is generally accepted

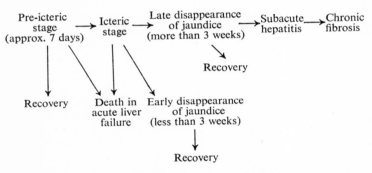

FIG. 10. Clinical course of acute hepatitis.

as the clinical course of acute hepatitis. The short pre-icteric stage of a few days may end in recovery without the development of overt jaundice or, rarely, may progress rapidly to acute hepatic necrosis and death before jaundice has time to develop. More frequently the pre-icteric is followed by the icteric phase, which varies greatly in its intensity and duration. On rare occasions acute necrosis may result in early death, but usually the icteric phase is followed within a short period by recovery. Sometimes it persists for ten weeks or more, and while recovery after this time frequently occurs, these patients very occasionally pass through a subacute stage terminating in a chronic fibrosis indistinguishable from cirrhosis of the liver.

Pre-icteric stage. During the pre-icteric phase there is bilirubinuria. A van den Bergh test performed on the plasma shows that the plasma bilirubin is mainly in conjugated form (even though the concentration of bilirubin is not raised above normal as shown by the positive direct reaction). A little later, but still during the pre-icteric stage, urobilin and urobilinogen appear in the urine. Serum enzymes, especially glutamic-pyruvate transaminase (see pp. 173–175), are increased. The precipitation and flocculation tests do not become positive until a later stage in the disease.

Short icteric stage. If the pre-icteric stage clears up we have acute hepatitis without jaundice, but as far as is known jaundice and hyperbilirubinæmia more usually supervene, the direct van den Bergh reaction remaining positive (conjugated bilirubin in the plasma). The quantities of bilirubin, urobilinogen and urobilin in the urine increase considerably, and the stools become clay-coloured. Plasma phosphatase appears to be increased at about the time when jaundice develops, or a little later, and a little later still the flocculation tests become positive. In many cases the jaundice clears up rapidly (i.e. within two or three weeks of its development). The serum enzymes return to normal at an early stage in these patients. Bilirubin usually disappears from the urine while hyperbilirubinæmia is still present, but urobilinogen and urobilin persist in the urine until jaundice has completely gone. Frequently the flocculation and precipitation tests, as well as the plasma phosphatase, remain abnormal, for some time after the clinical evidence of jaundice has vanished, but eventually become normal if complete recovery occurs.

Long icteric stage. Not all cases of acute hepatitis run this short course; in some patients jaundice becomes intense, and remains so for weeks. Those cases in which complete recovery ultimately occurs differ from those in which chronic fibrosis supervenes. There is usually intense jaundice, and little bile reaches the gut, so that urobilin and urobilinogen disappear from the urine. Flocculation tests are usually positive and plasma alkaline phosphatase is increased, but not to the levels observed in obstructive jaundice. The serum enzymes are increased, but sometimes not enough to be of value in differentiating the condition from obstructive jaundice. The diagnosis between these two diseases may, therefore, sometimes be difficult. An obstructive jaundice likely to be confused clinically with this stage of acute hepatitis is most often due to malignant obstruction. Repeated determinations of fæcal urobilinogen will reveal that in acute hepatitis significant traces of bile-pigment reach the gut. Where hepatitis lasts two months or more, as in cholestatic hepatitis, serum albumin tends to be low, and serum globulin to be increased. When recovery eventually occurs, urobilin and urobilinogen reappear in the urine and, in the absence of relapse, remain present until recovery is apparently complete. Pigment reappears in quantity in the fæces, and recovery then follows much the same pattern as in the milder forms of hepatitis already described, although the return of tests to normal may be delayed.

Acute necrosis of the liver. This condition is the most extreme form of liver damage. There is severe damage to large areas of the liver and destruction of the reticular framework. Death may occur from acute liver failure before jaundice develops. The urine contains urobilinogen and urobilin, as well as bilirubin. Tyrosine and leucine crystals are said to appear in the urine, but this is not common. Chromatography, however, reveals a gross amino-aciduria greater than is seen in normal urines. Plasma amino-acid nitrogen is increased from the normal range of 4–7 mg/100 ml. to values above 10 mg/100 ml., figures which are practically diagnostic. Many serum enzymes, especially the glutamic-pyruvate transaminase, are greatly increased in acute necrosis. There is often hypoglycæmia; later hypoalbuminæmia may develop, and hypoprothrombinæmia may result in bleeding from the mucous membranes uncontrollable by parenteral vitamin K therapy. Sometimes a fatal termination is reached within a few days, and then jaundice is often minimal.

Chronic hepatitis. In patients in whom the condition has progressed to chronic fibrosis after long and persistent, or relapsing, jaundice, the flocculation tests remain positive, often strongly so, and frequently the serum globulin is increased, sometimes up to 9 g/100 ml. Electrophoresis shows that this is mainly due to a great increase in the γ-globulin. The plasma phosphatase level may remain high, and often increases to levels usually associated with obstructive jaundice. Death may then occur either from hæmorrhage associated with portal hypertension or from intercurrent infection associated with ascites and œdema. Death from liver failure is less common, but when it does occur is heralded by the same features as acute liver failure in any other stage of hepatitis.

Just as the results of liver function tests change during the course of acute hepatitis, they may change during the course of chronic hepatitis. In particular, the results of the flocculation and precipitation tests and of electrophoresis may change during the course of the disease. The concentrations of serum enzymes will fluctuate, increasing whenever there is an active exacerbation of the diasease.

In post-hepatitis cirrhosis there is sometimes no clinical or biochemical evidence of abnormality, even though histological evidence of cirrhosis may be present. It is possible that cirrhosis, with portal hypertension and hepatic decompensation, can develop after a latent period of years during which all tests of liver function are apparently normal.

Cirrhosis of the liver. There appears to be no way of differentiating by biochemical tests a chronic fibrosis or cirrhosis after acute hepatitis from the condition which has apparently developed without hepatitis. Such cirrhosis may be in a symptomless stage, or may be associated with ascites or hæmatemesis or frank liver insufficiency comparable with that of acute necrosis. The last will be obvious clinically and will show abnormality by almost any of the hepatic function tests. The help of the biochemist may sometimes be required in the other forms of cirrhosis. The results of biochemical tests in cirrhosis may be extremely variable, presumably due to the different extents to which liver damage, circulatory impairment by fibrosis and liver regeneration play their part. Increases in the concentration of serum enzymes are particularly valuable and will parallel the severity and activity of the necrosis. A direct van den Bergh reaction may show the presence of conjugated bilirubin with or without hyperbilirubinæmia. Urobilinogen and urobilin may be present in the urine. The turbidity and flocculation tests are usually positive. The serum γ-globulin is frequently increased and the albumin decreased. The phosphatase in the serum may be above normal, and sometimes may reach levels usually encountered in obstructive jaundice. The excretion of coproporphyrin III is usually raised, but there is still discussion concerning the most satisfactory means of measuring porphyrin excretion. Plasma sodium and potassium may be decreased due to changes in anti-diuretic hormone and aldosterone in the body fluids (see Chapter III).

Jaundice in infancy. Jaundice in the newborn may be physiological or due to: (*a*) hæmolytic disease through the presence of maternal antibodies; (*b*) to congenital abnormalities of the biliary tract leading to obstructive jaundice; or (*c*) acute hepatitis.

During the first few days of life most infants show a mild but variable degree of jaundice. The plasma always gives a negative direct van den Bergh reaction, and therefore contains bilirubin itself and not the conjugated pigment. This was attributed to excessive destruction of red cells at this time, believed to be related to the correction of the polycythæmia consequent to the anoxia of intra-uterine life. It is, of course, well recognized that in the adult considerable increases in red cell breakdown must occur before hyperbilirubinæmia occurs, so great is the functional capacity of the liver. Even with severe hæmolysis, plasma bilirubin rarely exceeds 4 or 5 mg/100 ml., yet in the newborn this concentration is frequently attained, even though hæmolysis of the excess red cells must be minimal. Glucuronyl transferase, the enzyme responsible for conjugating bilirubin, is not synthesized in adequate quantity until shortly after birth, the deficiency being greater if birth is premature. During intrauterine life the deficiency appears to be of no consequence, bilirubin presumably being removed from the fœtel circulation via the placenta. At birth the infant's liver is required to take over this function, but is unable to do so until adequate amounts of the enzymes have been synthesized to permit conjugation of the bilirubin and its excretion into the bile. The physiological jaundice then disappears.

In hæmolytic disease of the newborn, due to the transplacental transfer of maternal antibodies formed in response to fœtal red cells which are incompatible with the plasma of the mother, jaundice is much more severe. The resulting hæmolysis leads to increased bilirubin formation, which even further exceeds the limited conjugating capacity of the immature liver. Jaundice may appear on the first day after birth, and plasma bilirubin may quickly increase to 20 mg/100 ml. or more. Such high concentrations which may also occur in premature babies may lead to kernicterus, with consequent permanent brain injury (see page 184), and exchange transfusions are essential to avoid this serious complication.

The diagnosis of obstructive jaundice due to congenital

obliteration of the bile ducts provides a problem, but in this condition the jaundice is due to accumulation of conjugated bilirubin (positive direct van den Bergh reaction), whereas in hæmolytic jaundice of the newborn the bilirubin is usually mainly unconjugated (negative direct van den Bergh reaction). However, after prolonged hæmolytic jaundice in the newborn conjugated bilirubin may accumulate in the blood, the direct diazo reaction will be positive and bilirubin may appear in the urine. Adult concentrations of γ-globulin are not present even in normal serum of infants, so that the turbidity reactions are useless in the diagnosis of hepatitis. Misleading results in the routine urine tests occur because reduction of bilirubin to urobilinogen occurs only when the gut has been fully colonized by bacteria, and this is often considerably delayed in the breast-fed baby. Serum transaminase estimations may be of especial value in the investigation of jaundice in infancy.

Jaundice due to drugs. Although jaundice occurring during therapy with monoamine oxidase inhibitors resembles acute hepatitis, that associated with the administration of chlorpromazine types of drugs and with certain steroids, e.g. methyltestosterone, norethandrolone, is cholestatic in type in resembling obstructive jaundice. The alkaline phosphatase is usually very high and the thymol turbidity tests are frequently normal. However, the γ-globulin fraction of the serum is frequently raised. The jaundice appears to be due to cholestasis and the diagnosis is not easy with the usual laboratory techniques. It is therefore essential that a history of administration of a drug should be excluded in all patients with the biochemical findings of obstructive jaundice. On the other hand, it must not be overlooked that patients receiving such drugs may suffer from surgical jaundice and biopsy or percutaneous cholangiography perhaps followed by laparotomy may be necessary.

Hepatic coma. Coma and other neurological complications may occur in acute necrosis, acute hepatitis bordering on acute necrosis and in some cases of cirrhosis of the liver. The precise mechanism of these neurological manifestations is still obscure. Hypoglycæmia and cholæmia cannot be responsible since hypoglycæmia is not a constant feature of the condition and hepatic coma is very rare in obstructive jaundice. Acetylcholine poisoning may also be ruled out, since the administration of cholinesterase inhibitors to patients in hepatic coma does not worsen the condition. There is, on the other hand, much evidence that the condition is due to ammonia intoxication. The ammonia is derived probably by bacterial breakdown of protein or other substances in the gut, is absorbed in the portal tract and normally removed by the liver, so that relatively small amounts pass into the peripheral circulation. When there is hepato-cellular failure, as in hepatitis, this ammonia passes through the liver and reaches the systemic circulation via the hepatic vein. In cirrhosis the blood from the portal tract is conveyed directly to the systemic circulation in the portal blood by the anastomotic circulation. It is possible that not ammonia itself but other substances, such as mercaptans and methionine sulphoxide, which are absorbed from the gut with the ammonia, may be of importance in the causation of hepatic coma.

The mechanism of the interference of cerebral metabolism by accumulation of ammonia in the blood is still unknown. Analysis of the ammonia content of cerebral arterial and venous blood has shown that in hepatic coma ammonia is taken up by the brain, and indeed the central nervous system tissue is unique in containing very large amounts of glutamic acid, which can combine with ammonia to form glutamine.

It has been suggested that the glutamic acid of the brain provides a mechanism whereby excess ammonia may be taken up and that when the available glutamic acid becomes lower still more ammonia is taken up by the conversion of α-ketoglutaric acid to glutamic acid. This would lead to an α-ketoglutaric acid deficiency, and since α-ketoglutaric acid is an important member of the Krebs cycle, by which carbohydrate, fat and some amino acid residues are oxidized, such a deficiency might interfere with cerebral function.

Further Reading

GRAY, C. H. 1961. "Bile Pigments in Health and Disease." Charles C. Thomas, Springfield, Ill.

SHERLOCK, S. 1968. "Diseases of the Liver and Biliary System." 4th ed. Blackwell Scientific Publications, Oxford.

SHERLOCK, S. 1966. "Biliary secretory failure in man: the problem of cholestasis." *Ann. intern. Med.* **65**, 397.

WILLIAMS, R. 1965. "The assessment of liver function." *Anæsthesia*, **20**, 3.

CHAPTER VI

THE PLASMA PROTEINS

In routine chemical pathology the concentration of total proteins in the body fluids may be determined by measuring either the content of total nitrogen or of a specific amino acid after precipitation of the protein by a suitable agent such as trichloracetic acid. Alternatively, some physical constant such as refractive index or specific gravity may be measured.

A knowledge of the concentration of total plasma protein alone is almost useless unless supplemented by the proportions of the various constituents of the plasma proteins. Formerly these were crudely separated by the addition of various concentrations of salts, of ethanol or of methanol, but moving boundary electrophoresis was long regarded as the most accurate method for separation and characterization. Zone electrophoresis on paper, cellulose acetate or on starch or other gel, while not giving accurate values for electrophoretic mobility such as are given by the moving boundary technique, permit useful separations of the plasma proteins into albumin, α_1- and α_2-globulins, β-globulin and γ-globulin in that order of mobility.

In zone electrophoresis a small amount (usually 0·05 ml.) of serum is placed near the top of a long strip of filter-paper or cellulose acetate soaked in a buffer of appropriate pH and ionic strength, and suspended in a closed tank. The two ends of the paper dip into large reservoirs of buffer to minimize electrolytic changes. The reservoirs are in turn connected to appropriate vessels containing electrodes. On applying a potential difference of 200 volts or more, the proteins migrate according to their isoelectric points and become separated along the length of the paper. Because of an endosmotic flow of medium in the reverse direction, the movement of the protein fractions is less than would otherwise occur; on this account the γ-globulin migrates towards the cathode (Fig. 11). By suitable staining and inspection of the stained strip rough estimations of the protein fractions may be obtained. More complete separations can be obtained if electrophoresis is performed in a solid gel in which

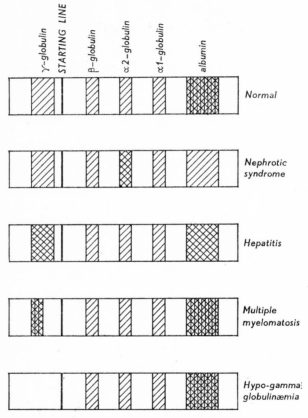

Fig. 11. Diagrammatic representation of paper electrophoresis of serum proteins in various diseases. The degree of shading roughly indicates amounts of protein fractions.

the protein solution on a piece of filter-paper is introduced into a slit. Photoelectric scanning is undesirable for routine work, since different proteins absorb stains unequally, and many problems of uneven migration, of flow unrelated to the electric field and of changes due to heating by the electrophoresis current require further investigation before significant quantitative results can be obtained. Nevertheless, if proper but time-consuming precautions are taken it is possible to obtain quantitative determinations of the various fractions of the plasma proteins. The fractions obtained can be identified from their positions, or in the case of gel electrophoresis by precipitation in the gel with specific immune sera, which may be prepared by the repeated injection of the protein fraction into a suitable animal.

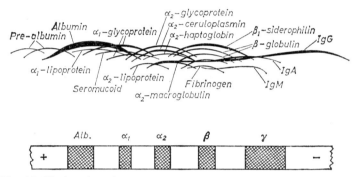

FIG. 12. The location of some of the identified serum proteins after immuno-electrophoresis in agar (*above*) and after electrophoresis in agar (*below*).

The antiserum is placed in a trough cut in the gel at the side of the electrophoresis pattern. Diffusion of the protein fraction and the antiserum will occur, and a line of precipitate will form where the two fractions meet. If the antiserum has been prepared against a crude mixture of proteins it will contain antibodies to all the proteins present, but if it has been prepared against a pure protein the antiserum will usually be highly specific for the purified protein. Such immune sera may be used in various ways for the qualitative and quantitative investigation of plasma proteins.

Immuno-electrophoresis using an antiserum prepared against whole human serum reveals many different proteins, some of which are illustrated in Fig. 11. Such a technique does not permit quantitative estimations of the various protein fractions, but this is now possible in a recently devised method, in which after separation by simple electrophoresis a second electrophoresis causes the separated proteins to pass into a sheet of gel containing anti-whole human serum antiserum. The proteins are precipitated with their specific antibodies, and the area under each protein peak is directly proportional to the concentration of that protein. In this way 45 distinct plasma proteins have been demonstrated, of which about 35 have been named and the function of 12 are known. Some are concerned with transport, e.g. cæruloplasmin (copper), haptoglobins* (hæmoglobin), α- and β-lipo-proteins, transferrins* (iron) and transcortins (cortisol). This technique may permit in the near future a greater understanding of the various alterations in the α- and β-globulins and other plasma proteins, the significance of changes in which have not been correlated specifically with any disease processes because of their heterogeneity.

Such techniques carried out with radioactively labelled proteins will permit measurements of the various protein pools and their rates of metabolism.

Protein-losing enteropathy. Some causes and effects of hypoalbuminæmia have been discussed elsewhere (page 40). The syndrome at first called idiopathic hypoproteinæmia, is now recognized as a protein-losing enteropathy. The plasma albumin concentration is low and accompanied by œdema, but there is no proteinuria or hepatic dysfunction or indication of malabsorption. The illness usually but not always presents in infancy or childhood. Œdema is usually the chief and only sign, and the fall in serum albumin was at first believed to be due to a deficient synthesis of albumin. However, ^{131}I labelled albumin, injected intravenously, disappeared from the circulation rapidly, but the ^{131}I label was much more quickly split from the albumin than had been expected, suggesting that there might be loss of intact

---

* Haptoglobins are α_2 globulins capable of binding free hæmoglobin and may play an important role in the degradation of hæmoglobin to bile pigments. They are destroyed during this process, and since the rate of their synthesis is slow, the concentration of haptoglobin falls in hæmolytic conditions. Different haptoglobins, the syntheses of which are clearly determined by allelic genes, occur in different populations, and there has been much study of their frequency and distribution. Similarly, there are a number of transferrins, the iron binding proteins, and these have also been found to be distributed differently in different populations. Such investigations are of considerable interest to the biochemist, geneticist and anthropologist.

albumin into the gut, where the ^{131}I would be liberated by digestive action. This was confirmed not only by the use of improved preparations of ^{131}I albumin but also with ^{131}I labelled polyvinylpyrrolidone (PVP) and ^{59}Fe-dextran; these last have molecular weights comparable with that of plasma albumin, but are not affected by the proteolytic enzymes in the gut. Protein loss by exudation into the gut has been shown to occur also in intestinal lymphogranulomatosis, gastric cancer, hypertrophic gastritis, regional enteritis, ulcerative colitis, the sprue syndrome, as well as some other gastro-intestinal disorders. In all but the last two conditions appropriate surgical resection can correct the loss. However, in true idiopathic hypoproteinuria the lesion seems to be associated with a congested œdematous intestine with a brown pigmentation in the external muscular coat. Studies with ^{131}I albumin showed that there is an abnormal transfer of protein across both surfaces of the gut, leading to an excessive exchange of protein into and out of the peritoneal cavity. Sometimes protein loss in gastro-intestinal conditions is also accompanied by severe deficiency in dietary calories. The combined protein-calorie deficiency is thus associated with a kwashiorkor-like syndrome with a severe illness with wasting and atrophy and loss of the skin and hair.

The γ-globulins. The normal γ-globulins are associated with antibody activity and from chromatography, electrophoresis, ultracentrifugation and immunological studies are known to be highly heterogeneous.

This heterogeneity is now accepted as a reflection of their polyclonal origin, i.e., of their formation from an enormously large number of immunologically competent cells, each of which can when stimulated by exposure to its own antigen, can proliferate to give clones of cells capable of producing corresponding antibody. The γ-globulins formed by these numerous clones of cells are known as the immunoglobulins. On electrophoresis the γ-globulins are not sharply separated from the β-globulins, the next fast-moving protein fraction, and separate into two overlapping groups, a faster γ_1 and a slower γ_2 group, each consisting of molecules with a relatively wide scatter of mobilities.

Immuno-electrophoresis using highly specific antisera prepared against purified fractions has shown three main classes of immunoglobulins, IgG, IgM and IgA, and more recently a further class, the IgD, has been described. IgD, a low-molecular-weight globulin composed of light B chains (see below), is present in normal serum and is normally excreted in minute quantity in the urine. The four main classes of immunoglobulins differ in their rates of synthesis, in their concentration in the blood plasma and in their distribution in the body fluids (Table 9).

Digestion of immunoglobulins with papain has revealed that all immunoglobulin molecules consist of polymers of greatly differing molecular weight but built up from similar basic structures consisting of two heavy A chains (molecular weight about 50,000) and two light B chains (molecular weight about 20,000) covalently linked by disulphide bonds. The light chains occur in two forms, K (κ) and L (λ), which are chemically and anti-

TABLE 9

Immuno-globulin.	Electro-phoretic mobility.	Concentration, mg/100 ml. plasma.	Rate of synthesis, G/day.	Body fluid with highest concentration.
IgG	a_2	1000	3	Extra-cellular fluid
IgM	a_1 or β_2	100	0·5	Blood plasma
IgA	a_1 or β_2	250	1·5	Secretions
IgD	? a_1	—	—	—

M refer to a macro-globulin; A refers to a fast-moving 7S compound of high carbohydrate content.

genically distinct and may be separated by electrophoresis into at least ten subfractions. The long A chains seem to be responsible for their ability to behave as an antigen.

The classification of the immunoglobulins and the chemical nature of the B chains has been assisted by the demonstration that the urinary proteins excreted in multiple myelomatosis (see pp. 90–92) cross-react immunologically with the B chains from normal human γ-globulin. An antiserum prepared in an animal against the Bence-Jones protein from urine will therefore precipitate with the purified B chains prepared by papain digestion of normal human γ-globulin.

Hypogammaglobulinæmia. This condition occurs in premature infants but may not be apparent until two months of age when it may be responsible for sudden overwhelming infection. The three immunoglobulins are synthesized at different rates but only after fœtal maturity. All infants whether mature or premature receive IgG across the placenta from the maternal circulation, and after birth this is metabolized and γ-globulin concentration may reach dangerously low levels within 6–8 weeks if the infant did not begin synthesis of his own, as will occur with the mature infant. In the premature infant (as opposed to the infant small because of malnourishment) the γ-globulin concentration reaches a dangerously low level by the time it has reached the weight it should have been at birth; it is still unable to synthesize immunoglobulins and will remain highly susceptible to infection until it is at that stage of development corresponding to a normal child of 8 weeks when it can synthesize its own antibodies. There is a case, therefore, for maintaining premature infants under sterile precautions until they have reached the stage at which they can synthesize their own antibodies; there is even a case for giving γ-globulin prophylacticly to all premature infants.

Hypogammaglobulinæmia, not associated with prematurity, is not rare and must be considered whenever there are unusually numerous bacterial or protozoal infections. Chronic bronchiectasis not responsive to orthodox treatment is common and some patients present as a rheumatoid arthritis, but many remain undiagnosed because of insufficient awareness of the condition. Diagnosis is important, for continued treatment with antibiotics or by transfusions of pooled γ-globulin is effective. Most cases seem to respond normally to virus infection, although a few infants with the conditions have died from vaccinia gangrenosa. Sensitive techniques, e.g., by immunoelectrophoresis, show that the γ-globulins are never completely absent and that the condition must be regarded more truly as a hypogammaglobulinæmia, rather than agammaglobulinæmia.

Kunkel's zinc sulphate test provides a simple screening test in that a completely negative result is suggestive of the condition, while a positive test definitely excludes the condition. The condition may be confirmed by paper electrophoresis and then preferably by immunoelectrophoresis with gel. The diagnosis should be made only after the age of six months, when the plasma of normal infants should contain 600–1,300 mg γ-globulin/100 ml. The diagnosis can be made when there is less than 200 mg/100 ml.

TABLE 10. HYPOGAMMAGLOBULINÆMIA
(After Kerr and Hall, 1966)

Type.	Plasma γ-globulin mg./100 ml.	Incidence.		Ætiology.	Mechanism.	Pathology of lymphoid tissue.
		Age.	Sex.			
Physiological (Transient)	150	2–6 months	M & F	Immaturity of antibody-forming mechanism	Decreased synthesis	Immature lymphoid follicles; plasma cells decreased
Congenital (a) Usual form	0–50	After 6 months	M	Genetic sex-linked recessive	Decreased synthesis	Lymph nodes devoid of normal follicles and plasma cells
(b) Alymphocytic form	0–50	3–8 months	M & F M predominates M & F	Non-sex-linked Genetic autosomal dominant; EARLY DEATH before 2 years of age	Decreased synthesis	Lymph nodes devoid of normal follicles and plasma cells plus thymic aplasia, no lymphocytes
Acquired (a) Idiopathic	0–200	Any age	M & F	? Atypical reticulosis or collagen disease	Decreased synthesis	Follicular hyper- or hypoplasia. Reticulum cell hypoplasia. Diminished plasma cells
(b) Secondary to disease of reticuloendothelial system	0–200	Any age	M & F	Thymic tumour, myeloma, Hodgkin's disease, chronic lymphatic leukæmia	Decreased synthesis	Granulomatosis or neoplastic infiltration, diminished plasma cells
(c) Secondary to globulin loss (1) Nephrotic syndrome	100–500	Any age	M & F	Various	Urine loss	Normal
(2) Idiopathic hypoproteinæmias	200–600	Any age	M & F	Gastric or intestinal diseases constrictive pericarditis	Increased loss into gut	Normal

The disease may be classified as shown in Table 10 which also shows the ætiology and pathology of the lymphoid tissues in this disease.

Hyperglobulinæmia. Raised serum globulins were recognized even with the earlier methods of investigation, and could be of diagnostic significance. If the total serum globulin was above 5·0 g/100 ml. the chances were 9 in 10 that multiple myelomatosis, sarcoid or a collagen disease was present. Values as high as 10–15 g/100 ml. occur in kala-azar. Fractionation of the plasma proteins has shown that even where the total protein is normal, there may be increased concentrations of globulin in many conditions. γ-globulins may be increased in cirrhosis and other diseases of the liver, in the reticuloses and sometimes in metastatic carcinoma. Hypergammaglobulinæmia has been said to precede all other symptoms of systemic lupus erythematosus. In a large group of infective diseases, including tubercle, syphilis and pneumonia, γ-globulin may be increased, but more usually there is an increase in one or more of the other globulin fractions. These may be increased in a miscellaneous group of diseases, including chronic renal disease, heart disease, diabetes mellitus, fractures, skin disease and old age.

Paraproteinæmias. This is the term applied to those conditions in which the plasma contains an unusual protein which in normal persons is either absent or present in such small quantity that it is not ordinarily detected. The term includes macroglobulinæmia, cryoglobulinæmia and multiple myelomatosis.

Macroglobulinæmia. This affects mainly the elderly and is characterized by the presence in the plasma of a macroglobulin apparently synthesized by lymphoid cells, infiltrating the marrow, lymph nodes and often other tissues. The plasma viscosity is increased and may result in vascular disturbances of the Raynaud type. There is reduced synthesis of antibodies and of the factors concerned with the clotting mechanisms. Sooner or later the disease becomes progressive with retinal changes, repeated epistaxis, an increased susceptibility to infection, hæmolytic anæmia and even neurological symptoms. The diagnosis should be considered in elderly people with rouleaux formation in the blood film and high sedimentation rate. Electrophoresis will reveal a band in the γ region and should be followed by immunological or ultracentrifugal examination.

Cryoglobulinæmia. Cryoglobulins are globulins of high molecular weight, mostly 7S, which precipitate spontaneously from the serum at low temperature.

Multiple myelomatosis. In multiple myelomatosis there is a profuse synthesis of protein, the nature of which can be very different in different patients. The disease affects the plasma cells and affects both serum globulin synthesis and antibody formation. Any given patient may have one or more of the following three abnormalities: (1) production of large amounts of myeloma globulin, an abnormal serum globulin characterized by electrophoretic migration as a very narrow intensely staining band rather than diffusely as normal γ-globulin; (2) excretion in the urine of Bence Jones protein, which differs from the myeloma and other globulins in being coagulated at 40–60° C and partial or complete solution at higher temperatures; (3) paramyloidosis, in which there are depositions of yet another abnormal protein in the tissues. There is no relationship between the clinical symptoms, the hæmatological aspect and the type of protein abnormality.

The disturbance of protein synthesis is profound. In some patients the concentration of abnormal globulin may be more than double that of the total normal serum protein; in these large amounts of protein not Bence Jones protein may be excreted in the urine; in others half of the total dietary nitrogen intake may be excreted as newly synthesized Bence Jones protein, the origin and function of which is as yet unknown. Bence Jones protein in the urine is almost pathognomonic, but is usually demonstrable in only half of the cases, although modern highly sensitive immunological methods may prove it to be present in all subjects. Hyperglobulinæmia, which is usually associated with rouleaux formation and a high erythrocyte sedimentation rate, is found in only 75 per cent of the cases, but in all an abnormal electrophoretic pattern is found. In some patients the circulating proteins are cryoglobulins.

The myeloma globulin which can migrate in the γ-, β- or even the α-globulin region is relatively homogeneous, whereas normal γ-globulins are a complex mixture. It is unknown whether the myeloma globulin, which is almost certainly synthesized in the tumour masses, is a new protein entity and due to a mutation or whether it represents an increased quantity of a normal but minor component of the plasma proteins. Chemical analysis of the amino acids has shown no uniquely characteristic composition.

Recently patients with essential hyperglobulinæmia have been described, but the protein is usually more heterogenous than the myeloma globulin. There are symptoms referable to the protein in the serum, but there are no skeletal changes, bone pain or histological picture of myeloma. Some patients have been found to develop true myeloma later and must be regarded as in a pre-myelomatous condition.

True Bence-Jones protein is probably diagnostic of multiple myeloma; the diagnosis was probably open to doubt in other diseases in which it has been described. This protein must be present in the plasma, but its concentration must be very low, because of the rapid rate of its excretion. The protein excreted by any one patient is usually the same throughout the disease, but the patterns by different patients vary enormously. Isotope experiments with ^{15}N or ^{14}C labelled amino acids show that there is extremely rapid synthesis of protein, and the rapidity of decline of the isotope content shows that there must be a synthesis *de novo* from amino acids and that the protein is not derived from circulating plasma proteins or tissue proteins. Bence-Jones protein is not a precursor of the myeloma globulin of the serum, but may well be a precursor or an abortive product of serum globulin synthesis.

Primary amyloidosis affecting the gastro-intestinal tract, the heart and blood vessels occurs not uncommonly in multiple myeloma. Those patients with both multiple myeloma and amyloidosis usually show Bence-Jones protein in the urine without hyperglobulinæmia.

Myeloma globulins and macroglobulins are believed to be due to autonomous neoplasia of lymphoid tissue consisting of clones of cells synthesizing large quantities of heavy A chains. Bence-Jones protein is therefore now believed to be due to such a faulty γ-globulin synthesis resulting in the liberation of free B chains which, because of their low molecular weight, are readily excreted in the urine. A corresponding over-production of heavy A chains has been described and is associated with changes in the lymph glands, but without skeletal lesions. Unlike multiple myeloma, an abnormal protein appears in the urine and in the plasma.

Table 11 summarizes the findings in the four main types of myelomatosis.

TABLE 11. TYPES OF MYELOMA

Immunological specificity of serum myeloma globulin.	Electrophoretic appearance.	Half life, days.	Bence-Jones proteinuria.	Renal failure.	Amyloid.
A chains of IgG	++++ prominent band	13	Usually less than 1 g/day* 30% L type B chain 70% K type B chain	Less common	Not uncommon
A chains of IgA	++ less prominent band	5	More than 1 g/day in 30% of patients 30% L type B chain 70% K type B chain	Common	Sometimes
A chains of IgD	Trace inconspicuous band	Short 2–4	*Always* more than 1 g/day ALL L type B chains	Very common	?
None detectable	—	—	Usually heavy	Very common	Sometimes

K = kappa; L = lambda. * Some of these may have no Bence–Jones protein.

Other abnormalities of the plasma proteins. Bisalbuminæmia due to an abnormal gene, is a very rare condition in which half of the serum albumin migrates abnormally during electrophoresis. Analbuminæmia is also a very rare abnormality, and there is now evidence that osmotic equilibrium is maintained by increases in many of the other constituents of the plasma proteins; œdema if present is therefore minimized in this condition.

C-reactive protein is a serum protein which can react with a somatic C-polysaccharide of the pneumococcus. It is absent from normal serum, but present in conditions associated with tissue inflammation or necrosis, such as in myocardial infarction.

In acatalasia there is a genetically determined deficiency of catalase in red cells, and the consequent accumulation of hydrogen peroxide in the blood can lead to a fatal condition, with gangrene of the superficial tissues.

Afibrinogenæmia is a cause of "obstetric accidents", and a deficiency of this protein occurs in liver disease and after major surgery, such as pulmonary resection (see p. 199).

FURTHER READING

"Medical Progress: The Gamma Globulins." 1966. *New Engl. J. Med.* **275**, 480, 652, 769.

Nobel Symposium 3. "Gamma Globulins." 1967. Interscience Publishers.

CHAPTER VII

CALCIUM AND PHOSPHORUS

Pathological calcification. Calcification may occur without the organization characteristic of bone formation. Such pathological calcification in tissues occurs not only in calculi but also in areas of dead tissue too large or so situated that they are not absorbed (e.g., in infarcts, in areas around organic foreign bodies such as the cysts of trichinella spiralis and echinococcus, in the caseous areas of tuberculous infections and in old inspissated collections of pus).

Calcification of masses of fibrous tissue occurs following hyalinization due to closing down of blood vessels, as in old scars and uterine fibromata. Calcification also takes place in areas of fatty degeneration found in atheroma and fat necrosis, and in the elastic laminæ of small arteries in the medial sclerosis of Monkeberg.

On the other hand, in metastatic calcification there is usually no obvious preliminary degeneration in the tissue before calcification; the condition is then associated either with the mobilization of calcium from bones, as in hyperparathyroidism, or as a result of excessive vitamin D therapy, when an abnormally high retention of dietary calcium may also play a part. Neoplastic calcification occurs in osteogenic sarcomata and osteomata. The mechanism of calcification in the rare disease calcinosis universalis is unknown. In these conditions calcium salts may be deposited in lungs, stomach, kidney, blood vessels and muscles.

Mechanisms of calcification. The salt deposited in calcified areas appears to be similar to that of bone, and the ions of which this material is composed must be brought via the bloodstream to the area undergoing calcification. Calcification has been alleged to occur by: (*a*) deposition of calcium phosphate due to simple alkalinization; (*b*) deposition due to super-saturation; (*c*) preliminary formation of calcium soaps followed by replacement of their fatty acids by inorganic acids; (*d*) liberation of inorganic phosphates by the action of phosphatase or phosphorylase; and (e) liberation of calcium from protein complexes by proteolytic enzymes. However, the precise mechanism remains obscure, although changes in citrate concentration may alter the complexing activity of tissues and lead to deposition of the bone salts.

Mechanism of bone formation. The calcium salt which is responsible for the rigidity of normal bone is continuously reabsorbed and new material deposited. This dynamic mechanism provides not only for growth when deposition is greater than reabsorption but also enables the bone structure to be highly organized and readily adaptable to the stresses to which it is exposed.

The bone salt is the so-called hydroxy apatite, $Ca_{10}(PO_4)_6(OH)_2$, but X-ray crystallography has shown that at the surface of the complex lattice of ions responsible for the crystal form there are defects which may be filled by sodium, magnesium, strontium or, in some circumstances, radium, as well as by carbonate, fluoride and citrate ions, which thus accounts for the variability in composition of the bone mineral.

Growth of bone occurs:

(a) at the growing end in a zone of provisional cartilage;
(b) at the periosteum, in which provisional cartilage formation does not take place;
(c) in the cortex and trabeculæ, providing a mechanism for the continuous remodelling of bone during growth.

This last—the endosteal bone formation—takes place not only during growth but continues throughout life, so that the bone is not a static structure, but is in dynamic equilibrium with the body fluids. In this way bone is constantly remodelled according to the stresses and strains to which it is subjected.

Bone formation takes place in two stages: by the laying down of an extracellular matrix—the osteoid—by the osteoblasts and the deposition of the bone salt in this osteoid. Side by side with this deposition of osteoid by osteoblastic activity and the deposition of calcium salts, osteoclasts destroy bone and liberate the ions of the bone-salts into the body fluids. Albright has represented this by a diagram similar to Fig. 13, where the inner square represents the skeleton in equilibrium with the body-fluids. On the right-hand side of this diagrammatic skeleton, osteoblasts are laying down osteoid, which in due course becomes calcified, so that the greater proportion of the skeleton is normal bone. On the left-hand side osteoclasts are destroying bone, liberating calcium and phosphorus ions which circulate in the body fluids, and are either re-utilized in calcification or

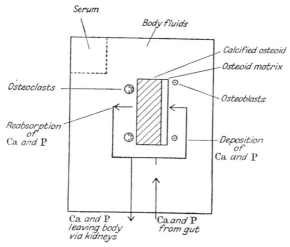

FIG. 13. Formation of normal bone (after Albright).

excreted via the kidney. Calcium and phosphorus are also shown to be entering the body via the gut. It is important to note that in the adult the calcium and phosphorus entering the body will precisely equal that leaving the body. The composition of the body-fluids with which the skeleton is in equilibrium may be investigated by analysis of serum or plasma.

The mechanism of calcification of the skeleton and the means by which it is under the control of hormones, vitamins and the numerous other unknown factors remains a major enigma of medicine. Bone crystals are so small that they present to the body fluids a surface area of between 50 and 250 sq. metres/g and since an adult skeleton may contain 2 kg or more of bone salt, calcification and decalcification can rapidly occur. Electrical and other changes occur at, near and just under the surface of these crystals, so that interstitial fluid of bone cannot be of the same ionic composition as the extracellular fluid of the body (Newman and Newman, 1958). Bone may therefore be decalcified, and calcium and phosphate can be absorbed from the intestine, even though plasma seems to be supersaturated with the bone mineral; on the other hand, new bone can be laid down, even though the immediate environment of the bone crystals is probably under-saturated.

Transport of calcium and phosphate between intestine and body fluid, and between body fluid and bone and renal tubular fluid, may be mediated by complexing agents such as citrate; local increases or decreases in citrate formation may be caused by vitamin D and parathyroid hormone interfering with certain stages of carbohydrate metabolism at this site of action.

The blood chemistry in bone disease

Serum calcium. Since the calcium of the circulating blood is confined to the plasma, determinations on whole blood are useless, for they would need to be interpreted in relation to the hæmatocrit reading. Also, since the commoner anticoagulants act by forming an insoluble or non-ionized calcium salt and would interfere with the determination, analysis for calcium must always be performed on serum.

The normal total serum calcium is maintained very constant between 8·5 and 11·5 mg/ml, and consists of three fractions:

(a) about 0·5 mg/100 ml. is diffusible but non-ionized;
(b) about 4·5–5·5 mg/100 ml. is non-ionized and is bound to protein, and therefore non-diffusible;
(c) about 4·0–5·5 mg/100 ml. is ionized.

Fraction (c), the ionized calcium, is important in maintaining normal neuro-muscular sensitivity probably by its effects on cell permeability to Na^+ and K^+ thus modifying polarization and depolarization. Methods of measuring this fraction have been described, but clinical investigations are usually confined to the total serum calcium.

Changes in total serum calcium due to changes in the ionized calcium are more important clinically, for a fall in serum calcium is then associated with tetany, and conversely, an increase is accompanied by diarrhœa and other disturbances of the gastrointestinal tract. On the other hand, decreased total serum calcium due to a fall in plasma protein concentration does not add significantly to the symptoms and signs of the primary disturbances of plasma protein. Hypercalcæmia due to hyperproteinæmia is rare because the albumin fraction is very rarely increased; most examples of hyperproteinæmia are due to increases in globulins, which are less important in combining with calcium. Hypercalcæmia, due to increase in the ionized fraction, is usually accompanied by hypercalciuria and diuresis, and may lead to metastatic calcification in tendons, muscles, and in the iris (keratitis) and kidney. Calcification in the kidney may lead to renal failure and death. If, then, an abnormal total serum calcium is found it is important to be certain that it is unaccompanied by a change in the total plasma protein. In classical hyper- or hypo-parathyroidism this may be obvious on clinical grounds, but when both calcium and protein deficiency exist side by side, as in disturbances of nutrition, such as pancreatic disease or the sprue syndrome, this may not be obvious, and determination of protein as well as calcium in the serum may be necessary. Tetany, either latent or overt, is always present when the ionized calcium is low; the condition also occurs in magnesium deficiency and in alkalosis, whether respiratory or metabolic in origin; the total serum calcium is then almost always normal. Tetany depends upon a state of hyper-excitability of the central nervous system; this is inversely proportional to the concentration of calcium, magnesium and hydrogen ions and proportional to the concentration of sodium and potassium ions.

Endocrinological control of calcium. The almost constant concentration of calcium in plasma has been considered to be due to parathormone secreted by the parathyroid glands. This hormone increases concentration of calcium in the plasma either secondarily to an effect upon phosphate reabsorption in the renal tubules or by a direct effect in mobilizing calcium from the bone. This hormone: (1) increases phosphate excretion in the urine by <u>decreasing reabsorption</u>; (2) decreases plasma phosphate; (3) increases plasma calcium; (4) increases urinary calcium excretion; (5) mobilizes calcium and phosphate from bone either secondarily to (1)–(4) or by direct action on bone; (6) increases calcium absorption from the gut. The effects on the kidney and gut are rapid, of limited capacity and sensitive to small changes in hormone; that in the bone is slower, of almost unlimited capacity and much less sensitive to the hormone. Stimulation and suppression of parathormone secretion probably occur in response to hypo- and hypercalcæmia respectively, thus providing a negative feedback mechanism.

A second hormone, thyrocalcitonin, is secreted by the thyroid, lowers plasma calcium rapidly and is probably of greater importance than the more slowly acting parathormone in maintaining the constancy of the plasma calcium. The recognition of this hypocalcæmic factor must modify our views concerning bone and other diseases of calcium metabolism. Thyrocalcitonin not only lowers plasma calcium but also increases phosphorus excretion, lowers plasma phosphorus and urinary hydroxyproline excretion; the last probably by inhibiting resorption of bone.

Clinically, thyrocalcitonin has been used in the treatment of idiopathic hypercalcæmia and that of malignancy. Much will depend upon the mode of action of this apparently important hormone, and upon the development of methods for measuring it and parathormone in blood.

Hypercalcæmia. A raised serum calcium is not common, but occurs following increased absorption as after over-dosage with calciferol (vitamin D) or following a prolonged diet of milk and alkali. The condition may also arise as a result of increased liberation of calcium from the bones, as in hyperparathyroidism, some cases of multiple myelomatosis, osteogenic sarcoma and secondary carcinomatosis. It occurs following immobilization, especially in childhood, and is found in sarcoid disease and idiopathic hypercalcæmia in infants.

Idiopathic hypercalcæmia in infancy is a not uncommon cause of ill health and failure to thrive. In its benign form there is anorexia, vomiting and constipation, but when severe there is physical and mental retardation, patchy osteoporosis, renal failure and hypertension. Balance studies show that there is abnormally high absorption and retention of calcium, and it has been considered possible that the condition is due to hypersensitivity to vitamin D. The condition, like hypercalcæmia due to sarcoidosis, responds favourably to a low calcium–high cereal diet or to adrenocortical steroids.

Hypercalciuria accompanies hypercalcæmia when renal function is good and is therefore usually found in all the circumstances leading to the latter condition. Hypercalciuria also occurs without hypercalcæmia in active osteoporosis whatever its cause, sometimes in Paget's disease and more frequently without obvious cause ("idiopathic hypercalciuria"). Sometimes this last condition may primarily be due to increased absorption of calcium from the gut—like hypercalcæmia due to sarcoidosis; on the other hand, it may be secondary to failure of renal tubular reabsorption.

Patients with hypercalciuria regularly excrete 400 mg/day or more of calcium in the urine; normally the urinary excretion is much less than this even when the calcium in the diet is high. Persistent hypercalciuria, whatever its cause, may lead to renal stone formation and even nephrocalcinosis. (The renal stones of renal tubular acidosis (see p. 18 and p. 25) are not always associated with hypercalciuria, but are due to the excretion of an alkaline urine in which calcium phosphate is virtually insoluble.) When the cause of hypercalciuria is known it must be removed or treated; when it is not it must be treated either by interfering with calcium absorption from the gut with sodium phytate or cellulose phosphate or by depressing renal tubular absorption of calcium with thiazadine diuretics.

Hypocalcæmia. Hypocalcæmia occurs with hypoproteinæmia in the nephrotic syndrome, protein deficiency, liver disease and chronic sepsis. The ionized fraction of the serum calcium is decreased in conditions associated with decreased absorption, such as in vitamin D deficiency, vitamin D resistance, as a result of phosphorus-rich diets, in renal rickets, in the sprue syndrome, in long-standing jaundice and sometimes following cation-exchange resins. Excessive urinary excretion of calcium may occur in renal tubular failure after long-continued use of mercurial and other diuretics. The commonest cause of hypocalcæmia is hypoparathyroidism, usually due to removal of parathyroid tissue at thyroidectomy. The condition usually necessitates treatment with large amounts of calciferol, as parathyroid hormone loses its biological effects on repeated injection. Prolonged disease of the ionized calcium causes cataract, probably because of the effect upon permeability in the lens.

Plasma phosphorus. In addition to its role in the formation of bone, phosphate plays an important part as an intracellular anion in energy exchanges, in the metabolism of carbohydrates and fats, and is an essential constituent of nucleoproteins, phosphoproteins and phospholipids. It is therefore not surprising that concentration in the plasma is affected by conditions other than bone disease. Thus a high plasma phosphate is observed in renal failure and also accompanies excessive secretion of growth hormone, as in acromegaly and childhood. The phosphate is depressed after insulin or a large carbohydrate meal. The fasting serum phosphorus in general varies inversely with the concentration of calcium, except when total serum calcium is abnormal because of changes in the protein-bound fraction due to hyper- or hypo-proteinæmia. Otherwise all conditions associated with high serum calcium are accompanied by low inorganic phosphorus, and all conditions associated with low serum calcium with a raised inorganic phosphorus. This does not occur in osteomalacia nor infantile rickets in which the product of the serum calcium and inorganic phosphate expressed in mg/100 ml. is less than 40, instead of between 40 and 60, the value found in normal infants.

Low concentrations of plasma phosphorus are characteristic of hyperparathyroidism or vitamin D deficiency associated with secondary hyperparathyroidism, and certain inborn errors of renal tubular function such as the Fanconi syndrome. In patients with these inborn errors there is a hereditary lesion of the renal tubule whereby substances such as amino acids, glucose and phosphate are not reabsorbed by the tubules. When phosphate is not reabsorbed there results a vitamin D-resistant rickets.

Phosphatases. Phosphatases are enzymes capable of liberating inorganic phosphate from organic phosphates. They are classified into acid and alkaline phosphatases, the former acting optimally at the acid pH of 5·1 and the latter at the alkaline pH of 9·5. They are estimated either by the quantity of phenol liberated from sodium phenylphosphate, as in the Armstrong–King method widely used in this country, or by the liberation of phosphate from glycerophosphate, as in the Bodansky method most frequently used in the United States. Alkaline phosphatases are found in bone, kidney, liver, bile, intestinal mucosa and plasma. Acid phosphatases are found in red cells and in prostatic tissue and in carcinomatous tissue derived from the prostate.

Plasma alkaline phosphatase. In both obstructive and hepatic jaundice the alkaline phosphatase of the plasma is increased above normal level of 4–13 units/100 ml., more so in the former (see Chapter V).

Alkaline phosphatase appears to play an important role in bone formation, and is usually regarded as a product of osteoblastic activity. Any condition associated with the laying down of new bone is therefore associated with an increase above normal in the plasma alkaline phosphatase activity. The increased osteoblastic activity due to rapid growth in infancy is reflected in an increase in plasma alkaline phosphatase to 15–20 units/100 ml. In general, osteolytic processes do not cause any alteration in the plasma alkaline phosphatase unless accompanied by an osteoblastic reaction. In Paget's disease, a condition in which excessive osteolytic and osteoblastic changes occur side by side, although there is no change in the serum calcium and plasma inorganic phosphate, the alkaline phosphatase is greatly increased, often to 60 or more units/100 ml. In hyperparathyroidism there is mobilization of calcium from the bone associated with increased serum calcium and decreased inorganic phosphate; osteoblastic activity tends to raise the plasma alkaline phosphatase. Some cases of multiple myelomatosis and osteogenic sarcoma present similar pictures. Simple tumours and secondary tumours in bone produce an increase in alkaline phosphatase only if there is a marked osteoblastic reaction, as occurs in carcinoma of the prostate and of the breast. In rickets the product of the calcium and inorganic phosphate concentrations is low, either because of low serum calcium or a low plasma inorganic phosphate. Osteoblastic activity is again very high, and the alkaline phosphatase of serum is greatly increased. The blood chemistry is normal in fragilitas ossium and in Albers-Schonberg disease.

Plasma acid phosphatase. Prostatic tissue contains an acid phosphatase which is normally secreted in the prostatic fluid. Carcinomatous tissue of prostatic origin can also secrete this acid phosphatase, and when this tissue is present in the form of metastases the acid phosphatase is secreted into the bloodstream; an increase of the plasma acid phosphatase above the normal level of 4 units/100 ml. is extremely important in the diagnosis of metastasizing carcinoma of the prostate. The plasma acid phosphatase which has been derived from the prostate is inactivated by treatment with alcohol. This property is sometimes used to decide whether an increase in the plasma acid phosphatase is due to prostatic carcinoma or to liberation of acid phosphatase from red cells by hæmolysis. Prostatic carcinoma commonly metastasizes in bone, and usually results in a considerable osteoblastic reaction in the surrounding bone. Accordingly, there is a considerable increase in both the acid and alkaline phosphatase. When œstrogens are administered the acid phosphatase rapidly falls to normal concentrations, but the alkaline phosphatase increases for a week or two, and then during a period of months gradually returns to normal. It seems likely that the rapid disappearance of acid phosphatase is due to the regression of the carcinoma metastases and that the osteoblastic reaction after a short period of intensification takes several months to disappear completely. This intensification of the osteoblastic reaction, which may in part be due to filling in spaces left by the regressing tumour, may account for the increased density of bone as seen by X-ray after œstrogen therapy, which contrasts with the striking symptomatic improvement. Very rarely prostatic carcinoma in bone is almost purely osteolytic, and then the acid phosphatase only is increased above normal, the alkaline phosphatase remaining within normal limits.

Classification of bone diseases

Albright has classified general metabolic disease of bone as osteoporosis, osteomalacia and hyperparathyroidism.

Osteoporosis. Osteoporosis may occur:

(*a*) after disuse;
(*b*) in extreme old age;
(*c*) in malnutrition, particularly when protein is deficient;
(*d*) in thyrotoxicosis;
(*e*) in Cushing's syndrome and during treatment with adrenocortical steroids;
(*f*) occasionally after the menopause, when it is believed to be related to the deficiency of œstrogen found at that time.

There is also an idiopathic osteoporosis of unknown ætiology. In all these conditions there appears to be a deficiency of osteoid formation and there is a general decrease in the mass of bone apparently related to a primary hypoplasia of the osteoblasts. The calcification of the osteoid present is normal, but the reduced mass of bone tissue causes a decreased requirement of calcium and phosphorus, which are therefore excreted in increased quantities in the urine. Recently, however, Nordin and Frazer have claimed that in spite of the general belief that the loss of bone salts is secondary to deficiency of osteoid, some patients with osteoporosis respond favourably to a long-continued high cal-

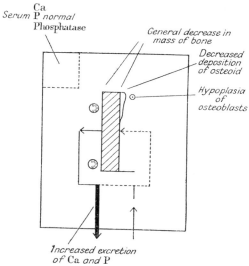

FIG. 14. Formation of bone in osteoporosis (after Albright).

cium diet. They state that in experimental animals osteoporosis results from calcium deficiency alone, whereas osteomalacia results when there also is vitamin-D deficiency. Osteoporosis has been represented by Albright as shown in Fig. 14. Calcium, phosphorus and phosphatase concentrations in the body-fluids are perfectly normal.

Osteomalacia. Osteomalacia is due to a deficiency of calcium and phosphorus in the body. It may be due to:

(a) a deficiency of dietary vitamin D or of calcium*;
(b) the diminished absorption of calcium associated with the sprue syndrome, regional ileitis or intestinal shunts;

* In Britain osteomalacia due to dietary deficiency alone is rare and is usually precipitated by excessive loss of calcium as in pregnancy or lactation.

(c) hyperchloræmic acidosis (see page 26);
(d) after ureteric transplants (see page 26);
(e) the increased excretion of phosphorus via the gut in renal rickets.

The last condition occurs in growing children with chronic renal disease. The kidneys fail to eliminate phosphorus, which is then eliminated via the gut. The increased quantity of phos-

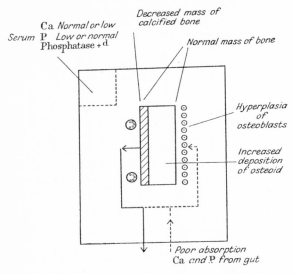

FIG. 15. Formation of bone in osteomalacia (after Albright).

phorus in the gut is said to impede the absorption of calcium. In osteomalacia the mass of bone is normal or increased (as shown by the swelling of the epiphyses in rickets), but the calcification of that bone is poor. Areas of uncalcified osteoid may occur in bands giving the appearance of fractures (pseudo-fractures) or in zones (Loeser's zones). There appears to be hyperplasia of osteoblasts and an increased deposition of osteoid, which is inadequately calcified because of the reduced amounts of available calcium and phosphate (see Fig. 15). In osteomalacia of most kinds the serum calcium is low, as well as the phosphorus and the phosphatase greatly increased. There is a tendency for secondary hyperparathyroidism to restore the serum calcium to normal and at the same time lower the plasma phosphorus.

Osteomalacia or rickets also occurs in a variety of relatively rare conditions, all characterized by an increased urinary excretion of phosphate. In the Fanconi syndrome (see Chapter IX) this is often associated with renal glycosuria, amino aciduria and sometimes with an inability to excrete an acid urine.

Hyperparathyroidism. In hyperparathyroidism the excess of circulating parathyroid hormone is believed not only to mobilize calcium and inorganic phosphate from the bones but also to increase reabsorption of phosphate in the renal tubules. The increased excretion of phosphate in the urine diminishes the concentration of phosphate in the plasma and is said to cause an increased mobilization of calcium from the bone. The serum calcium becomes raised and there is increased excretion of calcium in the urine. The resulting decalcification is accompanied by increased osteoblastic activity, and the general weakening of the bone is believed to stimulate osteoblastic activity so that the dynamic equilibrium in the bone is greatly accelerated (Fig. 16). The increased osteoblastic activity is usually accompanied by very high alkaline phosphatase concentrations in the plasma.

Hyperparathyroidism, which may be due to either hyperplasia or tumour of the parathyroid gland, usually leads to generalized fibrocystic disease of bone, to be distinguished from local fibrocystic disease of bone, in which a similar change in only one part of the skeleton is not associated with abnormality of the parathyroid gland. There is also disseminated fibrocystic disease of bone (or polyostotic fibrous dysplasia), in which areas of fibrocystic disease are separated by normal bone, and this condition, too, is not related to hyperparathyroid over-activity. Whereas in hyperparathyroidism the serum calcium and alkaline phosphatase are usually raised, and the phosphate low, in these other conditions the blood chemistry is often normal, except for a raised alkaline phosphatase. Sometimes decalcification occurs in hyperparathyroidism without cyst formation or the excessive loss of calcium is balanced by a high intake; no bony lesions then occur, and the symptoms are referable entirely to the urinary tract, i,e., renal stone, nephrocalcinosis. In such cases the blood chemistry may frequently be normal.

Diagnosis of hyperparathyroidism. About 4 per cent of patients with renal stones and about 15 per cent of those with recurrent renal stones have hyperparathyroidism. When this condition is unassociated with the typical skeletal abnormality or the usual disturbances of serum calcium, phosphorus and phosphatase, the diagnosis may be very difficult, and if the condition has been

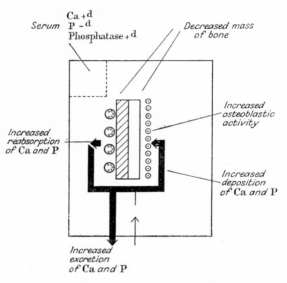

FIG. 16. Formation of bone in hyperparathyroidism (after Albright).

allowed to progress until there is secondary renal failure, the differentiation from primary renal failure with secondary hyperparathyroidism may be impossible. Since the values may be only intermittently raised, the serum calcium should be measured at monthly intervals for some months or even years in patients who might have hyperparathyroidism. Even a single raised value may be of diagnostic importance, providing the method used is accurate. As a screening test some workers estimate the urinary calcium excretion on a low calcium diet (say, 100–150 mg/day), and if the daily calcium excretion is much higher than about 150 mg follow this with a proper calcium balance test (see below). This may be of value if there is normal renal function, but some patients without hyperparathyroidism have hypercalciuria, while others with hyperparathyroidism may excrete low or normal amounts of calcium. Some workers have claimed that the ratio of phosphate clearance to that of creatinine is of diagnostic value, since one of the effects of parathyroid hormone is to decrease tubular absorption of phosphate, but in one group of 77 patients with hyperparathyroidism 39 per cent did not give an abnormal result. Prunty and his colleagues inject intravenously a progressively increasing amount of buffered sodium phosphate solution, and are thus able to measure urinary excretion of phosphorus at different concentrations in the plasma. When the urinary phosphorus in mg/min. is plotted against the serum phosphorus, the line obtained on extrapolating will cut the abscissa at the renal threshold. This test is not diagnostic, since the low threshold for phosphorus is also found in certain hereditary defects of tubular function. Rose has claimed that the ionized fraction of the serum calcium is raised in patients with primary hyperparathyroidism, but others find that this fraction is increased only when the total calcium is raised. Work continues on this difficult diagnostic problem, and a calcium or strontium infusion, perhaps with radioactive isotopes (^{45}Ca, ^{32}P or ^{80}Sr), may prove necessary for the diagnosis to be made in difficult subjects. Others have suggested that the failure of adrenocorticosteroid to decrease a raised serum calcium is highly suggestive of hyperparathyroidism. The immunological estimation of parathormone, although practicable, is not yet available for routine use.

Paget's disease, In Paget's disease there is grossly increased osteoblastic and osteoclastic activity leading to a complete disorganization and remodelling of the bone structure. Serum calcium and plasma phosphate are normal, but the high osteoblastic activity leads to a considerable increase in the plasma alkaline phosphatase, the height of which is dependent upon the extent of the disease. In the generalized form the alkaline phosphatase may be very high and exceed values of 100 units/100 ml. Indeed, the increase in phosphatase per unit of bone affected is greater than in any other bone disease. Owing to the increased vascularity of the bone, there is a greatly increased blood flow and cardiac output, which can readily lead to cardiac failure.

Calcium balance. A calcium balance investigation may be necessary for the diagnosis of hyperparathyroidism or to establish the effect of therapy on calcium absorption. The patient is fed a diet of constant and known calcium content (about 100 mg/day in hyperparathyroidism). After several days the calcium contents of the urine and fæces are estimated over a period of at least 12 days. Chromium sesquioxide, which is green and not absorbed, may be used as a marker to avoid error in timing the collections of fæces due to irregular bowel habits. The changes of calcium balance are best recorded in the form of a chart in which intake is plotted below the zero line and the combined urine and fæces excretion plotted above the line representing the intake. If the resulting plot is above the zero

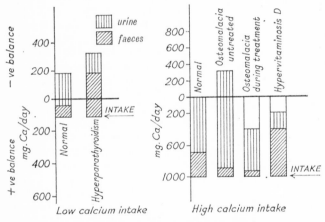

FIG. 17. Results of calcium balance experiments.

line output exceeds intake and the calcium balance is negative; if it is below the zero line intake exceeds output and the calcium balance is positive. Fig. 17 shows the results of two typical calcium balance investigations. These tests can be done only when special facilities are available.

Further Reading

ALBRIGHT, C. and REIFENSTEIN, E. C. 1948. "The Parathyroid Glands and Metabolic Bone Disease." Williams and Wilkins Co., Baltimore.

"Bone Metabolism in Relation to Clinical Medicine." 1963. Edited by H. A. Sissons. Pitman Medical; London.

FOURMAN, P. and ROYER, P. 1968. "Calcium Metabolism and the bone." 2nd Edition. Blackwell Scientific Publications: Oxford and Edinburgh.

JACKSON, W. P. U. 1967. "Calcium Metabolism and Bone Disease." Edward Arnold; London.

CHAPTER VIII

THE GASTRO-INTESTINAL TRACT

Gastric analysis. Analysis of gastric contents removed from a patient through a gastric tube might be expected to provide valuable information concerning gastric function. This is not so for gastric secretion varies greatly not only between individuals but is also influenced by local disease of the stomach, by reflex activity due to disease of other parts of the alimentary canal such as the appendix or gall-bladder, by general disease, as well as by psychogenic factors due to strange surroundings and general distress caused by swallowing the tube.

Although analysis of samples of gastric contents at various times after a test meal of alcohol, gruel or toast might be expected to be of clinical interest in providing curves of gastric secretion in response to the stimulus of food, in fact one only obtains a rough indication of the way in which the meal may be diluted by salivary, gastric and duodenal secretion. The curves obtained used to be classified into normal, achlorhydric and hyperchlorhydric curves according to the highest concentration of free hydrochloric acid attained. Examination for food residues revealed the emptying time of the stomach under the conditions of the test. In pyloric spasm there is persistent hyperchlorhydria and there is delayed emptying; the volume of gastric residuum is not so high as in pyloric stenosis, in which a plateau curve of free acid is obtained and food residues are present in the gastric residuum. Fractional test-meal curves obtained in practice are by no means so easily classified. Despite requests to the contrary, patients swallow their saliva, which with regurgitation of duodenal contents produce irregularities in what might have been smooth and regular curves. Little useful information is gained from these tests, which cannot be obtained by simpler and less time-consuming investigations. In research useful information is obtained by leaving a stomach tube in position for a day or more and continuously recording the pH while the ordinary meals are ingested.

Analysis of gastric contents is now performed only as part of the following tests:

1. Examination of the gastric residuum.
2. The histamine test in its various forms.
3. The insulin test.

The most useful routine investigation is either the examination of the gastric residuum, followed by the histamine test if preliminary tests on the gastric residuum reveal the absence of free hydrochloric acid. In all cases the following features are noted:

1. The volume.
2. The colour may be yellow or green because of regurgitation of bile, "coffee-ground" because of hæmatin derived from blood in an acid juice, or bright red because of recent hæmorrhage into a neutral juice.
3. Presence of mucus (only of importance if patient has not swallowed saliva).
4. Smell. It may be foul because of bacterial putrefaction.
5. Food residues. Starch (shown by positive iodine test) may be present from the previous meal, or there may be remnants of a charcoal biscuit given the night before.
6. Reaction to litmus, and if acid, to Congo Red paper.

Finally, each specimen obtained is analysed for free and total HCl by titrating to pH 2·5 and 8·5, using either thymol blue with its double colour change (1·2–2·8 and 8·0–9·6) or Topfer's reagent (2·9–4·2) and phenolphthalein (8·3–10·0).

Although methods are available for the quantitative determination of enzyme activity—e.g., of pepsin and rennin—in gastric secretions, they have never been adopted as routine, and the chemical pathologist is only rarely required to report on the presence or absence of these enzymes.

Free acid in gastric juice

Hyperchlorhydria is present when the free hydrochloric acid in the gastric juice exceeds 55 mEq/l.

Isochlorhydria is present when the free hydrochloric acid in the gastric juice is between 20 and 55 mEq/l.

Hypochlorhydria is present when the free hydrochloric acid in the gastric juice is less than 20 mEq/l.

Achlorhydria is present when the gastric juice contains no free hydrochloric acid.

Except when there is bacterial putrefaction due to malignant obstruction, the total acid is not more than 15–25 mEq/l. greater than the free acid.

The gastric residuum. This is the simplest test of the rate of gastric emptying. The patient eats a charcoal biscuit in the evening before and after fasting overnight the entire stomach contents are removed by stomach tube.

The volume of the normal gastric residuum usually amounts to 20–100 ml., but may vary from 0 to 120 ml. In pyloric obstruction, whether due to pyloric spasm or stenosis or malignant obstruction, the volume is greater than 150 ml., and in these last two conditions often exceeds 500 ml. When obstruction is due to pyloric stenosis food residues are usually present from the day before, but there is no putrefaction, because there is either isochlorhydria or, more usually, hyperchlorhydria. When obstruction is due to an advanced malignant growth the food residues are mixed with blood, and achlorhydria is almost always present. Bacterial contamination and putrefaction can occur, and there is a foul smell. With the exception of pyloric stenosis and malignant obstruction, it is impossible to diagnose any disease by an examination of the gastric residuum. However, the gastric residuum is hyperchlorhydric in approximately 50 per cent of cases of peptic ulcer and in only about 10 per cent of normal persons. Blood in the stomach is rapidly converted into acid hæmatin by the hydrochloric acid, and then looks like "coffee grounds". The presence of minute traces of blood in the gastric residuum is more often than not due to trauma by the stomach tube and of no pathological significance, but such an interpretation must not be too readily accepted in a person suspected of having an early carcinoma of the stomach. It must be remembered, too, that blood in the stomach may be due not only to local disease of the stomach but also to general bleeding disease, such as hypoprothrombinæmia, hæmophilia or leukæmia, as well as to œsophageal varices which may be present in chronic liver disease. It is obvious, therefore, that the results of analysis of the gastric residuum provide evidence of very limited value, the significance of which requires to be assessed together with the clinical findings and other investigations.

If a hypochlorhydric or achlorhydric gastric residuum is found its significance cannot be assessed until the response to histamine has been determined. Since histamine provides a very powerful stimulus to gastric secretion, such cases should always have the histamine test, and if the free acid increases to values above the limit of hypochlorhydria gastric secretion may be regarded as normal.

The histamine test. In the histamine test the removal of the gastric residuum is followed by the subcutaneous injection of 0·25–1·0 mg histamine (caution is necessary with patients suffering from asthma and other forms of sensitivity). Half an hour after the histamine injection the stomach contents are again removed and analysed. The histamine test is of particular value when it is desired to know the maximum concentration of acid which the stomach is capable of producing, especially in the investigation of pernicious anæmia, hypochromic anæmia and certain forms of simple dyspepsia. Pernicious anæmia is always associated with a histamine-fast achlorhydria, whereas this is not the case with other forms of megaloblastic anæmia. Carcinoma of the stomach, even in its early stages, is often associated with such an absolute achlorhydria or hypochlorhydria. With these exceptions, the gastric residuum provides as much information as the histamine test. In circumstances in which the free acid can be determined immediately after withdrawal of the gastric residuum the response to histamine often need not be determined unless there is achlorhydria or hypochlorhydria. Nevertheless, it must be remembered that about 4 per cent of normal individuals are achlorhydric, and whether this is temporary or whether these persons are more likely to develop pernicious anæmia has not yet been determined.

Gastric function is best investigated using the augmented histamine test, in which after a suitable antihistamine (e.g., 50–100 mg mepyramine maleate) gastric secretion is maximally stimulated by a larger dose of histamine (0·04 mg histamine acid phosphate per kg body weight). If a small volume of gastric juice is expected, as in achlorhydria, it is an advantage to use a radio-opaque tube and to screen it into position in the most dependent part of the stomach.

Recently, measurements of pH of gastric juice instead of titrations have been used for assessing acidity. Free hydrochloric acid is present when the pH is less than 3·5; achlorhydria is present when no specimen has a pH less than 3·5 and the maximum pH difference between the fasting juice and the most acid specimen after histamine does not exceed one pH unit. Hypochlorhydria is present when no specimen has a pH less than 3·5 but the maximum pH change after histamine exceeds one pH unit. The presence of bile invalidates the test unless the pH is in any case less than 3·5.

The insulin gastric function test. After first removing the gastric residuum 20 units of insulin are administered subcutaneously and 10 ml. of gastric juice are removed every 15 min. until the stomach is empty. The effectiveness of the insulin in lowering blood sugar should be confirmed by analysis. The test has been used mainly to assess the completeness of vagotomy, the test being performed before and after the operation.

The tubeless dye test. This provides a valuable screening test which may save the patient the unpleasantness of swallowing a stomach tube. A cation-exchange resin, in which the acidic centres have been replaced by a suitable dye, is given by mouth after the bladder has been emptied. If free hydrochloric acid is present in the gastric juice the hydrogen ions of this exchange with the dye in the resin. The liberated dye is then absorbed into the bloodstream and is excreted in the urine, in which it is readily detected by inspection and estimated colorimetrically. If achlorhydria is present very little of the dye is liberated and only minimal quantities appear in the urine. If necessary, gastric secretion can be stimulated a short time after the resin has been given, either by an injection of histamine or by a tablet containing caffeine sodium benzoate. The appearance of significant quantities of dye in the urine may safely be accepted as showing the presence of free acid in the gastric juice, but false negatives may sometimes be obtained, so that a negative result must always be confirmed by a proper examination of the gastric juice after histamine. Quantitative estimation of the dye can make the test more accurate in borderline cases.

Recent trends. In special centres more elaborate methods of investigating gastric function are being developed, but their precise value in diagnosis has yet to be assessed. Tests are required to assess the ability of the stomach to secrete acid, pepsin and intrinsic factor. Acid secretion is best estimated from the pH of gastric samples taken at frequent intervals (or by measuring continuously the pH of the gastric contents by a suitable electrode system) throughout the night and day while the patient is taking normal food. If the pH of any sample is below 3·5 acid has been secreted.

The rate of secretion of acid may be estimated by intubation (with radiological control of the position of the tube) and continuous aspiration. The volume, pH and acidity (as measured by titration to pH 7 either electrometrically or with phenol red as indicator) are measured for each sample collected under fasting conditions and after a maximum stimulus with histamine. The acid secretion may be expressed in mEq/hr. (Table 12).

TABLE 12—NORMAL VALUES OF ACID SECRETORY RATE
(Baron, 1963)

(a) *Basal secretion*
Males   1·3 ± 1·6 mEq/hr. (upper limit 4·5)
Females 1·1 ± 1·75 mEq/hr. (upper limit 4·6)

(b) *After histamine*
Males   0–60 min.   17·1 ± 11·9 mEq. (upper limit 50)
        Peak half hr. 10·8 ± 6·9 mEq. (upper limit 25)
Females 0–60 min.   9·4 ± 7·2 mEq. (upper limit 24)
        Peak half hr.   6·1 ± 4·5 mEq. (upper limit 15)

These figures apply to all normal subjects in London. Other normal values have been published in other localities.

Anacidity is defined as present when the pH never falls below 6 and is present in steatorrhea, hypochromic anæmia as well as pernicious anæmia. Free acid, combined acid, total acid, achlorhydria, hypochlorhydria and hyperchlorhydria are terms which are avoided in special research centres. Gastric ulcer is often associated with a low secretory rate, but anacidity is never present with a non-malignant ulcer. Gastric carcinoma is usually associated with anacidity and evidence of ulceration. An acid secretory rate under maximum stimulation of greater than 20 mEq/hr. is suggestive but not conclusive evidence against a diagnosis of gastric cancer. An output above the normal range is evidence in favour of duodenal ulceration, but a stimulated acid secretory rate of less than 10 mEq/hr. is evidence against a duodenal ulcer. Stomal ulcers after partial gastrectomy provide a special problem and may be associated with a ratio of unstimulated to stimulated secretion of more than 2:5. This is also characteristic of the Zollinger–Ellison syndrome, which is also characterized by high acid secretion under unstimulated conditions. This condition is associated with islet cell tumours of the pancreas, and can be diagnosed with certainty on gastric function tests of this kind.

Estimations of serum and urine pepsinogen provide an indirect measure of the peptic cell mass and may assist with the diagnosis of pernicious anæmia. The ability of the stomach to secrete intrinsic factor may be assessed by measuring hepatic uptake of radioactively labelled vitamin B_{12} by external counting over the liver area. When absorption is found to be impaired, the tests should be repeated giving intrinsic factor with the radioactive vitamin B_{12}. Alternatively urinary excretion of radioactivity or estimation of unabsorbed vitamin in the fæces may be used.

Pancreatic diseases. Acute pancreatitis presents as a surgical emergency, and chemical determination of the urinary and/or plasma amylase may be of great value in assisting diagnosis. These determinations should be performed as soon as possible. High concentrations in the urine and the plasma are almost diagnostic, but are also encountered in mumps and for a short period following perforation of duodenal ulcer. If this last condition is likely on clinical grounds it may be useful to repeat the amylase determination a few hours later. The interpretation of values for amylase concentrations depends upon the method used. In the author's experience the Somogyi method is best for plasma, and the upper limit of normality is 180 units/100 ml. For urine the old-fashioned Wohlgemuth method is more convenient, and the upper limit of normality is 35 units/ml.

Chronic pancreatitis is more difficult to diagnose. Very rarely there may be an increase in the urinary amylase, but the diurnal variations in the amylase excretion are so great that 24-hour specimens must be examined. Figures above 35 units/ml. or 50,000 units/day support the clinical diagnosis of chronic pancreatitis. More substantial evidence is provided by microscopy of the fæces, when fat globules, undigested meat fibres and starch granules may be observed. The condition is considered in greater detail below, where an account is also given of congenital fibrocystic disease. This is an important cause of failure to thrive during the first year of life, and early diagnosis is important, for treatment is beneficial only when started early. Amylase estimations are of no value in carcinoma of the pancreas.

Small intestine function. The absorptive capacity of the small intestine for protein can be assessed by measuring the daily fæcal output of nitrogen while the patient is ingesting a constant known amount of protein. Absorption, which is normally rapid, may also be determined using isotopically labelled protein, protein hydrolysates or amino acids.

Absorption of fat. Fats are probably absorbed in two ways: (1) a small proportion is absorbed into the portal system as hydrolysed fat and passes to the liver, and (2) a large proportion is absorbed into the lymphatics as unhydrolysed fat which has been emulsified by the partially hydrolysed glyceride fatty acid–bile salt complex. Support for this view is provided by the observation that when oleic acid—i.e., hydrolysed fat—is given by mouth the fat content of the peripheral blood hardly changes, presumably because it is all absorbed as such by the portal system and passes to the liver. On the other hand, when olive oil—i.e., unhydrolysed fat—is given by mouth there is a considerable increase in the fatty content of the systemic blood, presumably because this is mainly absorbed as such via the lymphatics and passes via the thoracic duct to the general circulation.

Defective absorption of fat usually leads to steatorrhœa characterized by the passage of fluid or semi-fluid, bulky pale, fatty and usually offensive stools. Their volume may amount to 500—1,000 ml. instead of 100–200 ml. They are more fluid than usual because of the poor water absorption consequent upon the rapid transit of intestinal contents, which may often be due to the production of short-chain fatty acids by bacterial digestion of carbohydrate. They also contain more solids than usual, on account of the extra fat and increased number of bacteria. The estimation of total bile pigment shows that the characteristic pale or putty colour is not due to reduction of these pigments but is more likely due to the reducing flora converting the bile pigments to their chromogens. A significant proportion of the fæcal fat is not of dietary origin, but at present this is of little practical importance, except when subjects are on low fat diets. This non-dietary component of the fæcal fat may well arise from either bacterial synthesis or from actual secretions of the intestinal tract. The fats may be present as long-chain fatty acids or their soaps or as triglycerides, and if these are present in excess may be recognized microscopically by the sheaves of crystals, copper sulphate-staining plaques or round refractile droplets respectively. Short-chain fatty acids formed by carbohydrate fermentation may also be present.

The estimation of fæcal fat. It is now generally recognized that estimations of the fat content of the dried stool are quite useless in the diagnosis of steatorrhœa. The condition should be diagnosed and its severity assessed by estimating the daily fæcal output of fat, preferably while the patient is receiving a diet containing a known amount of fat. Because of the great variation in regularity and size of stool, it is necessary that the investigation should be carried out over several days. Recently an improved technique has been devised which permits the estimation of the fat content of the stool to be determined each day. The mean fat content of the stools on the 1st, 2nd and 3rd, 2nd, 3rd, 4th, 3rd, 4th and 5th days, etc., is calculated and the running means thus obtained permit rapid assessment of the significance of the estimations. With a daily intake of 50 g of fat the absorption should be at least 90 per cent. This is time consuming, and many laboratories prefer to estimate simply the daily fat content of the fæces. On a normal diet the total amount of fæcal fat should not exceed 5 g/day.

Absorption defects. Steatorrhœa or fatty diarrhœa may be associated with a number of conditions associated with gross irreversible pathological changes in the intestinal wall. Thus it can occur after operative removal of the small intestine, with the development of intestinal fistulæ, with an intestinal loop syndrome, with tuberculous and regional ileitis, in scleroderma, lymphatic obstruction, in diabetes and following irradiation. In all of these conditions the treatment is that of the primary condition supplemented by a low fat–high protein diet. The carbohydrates should be controlled and there should be plenty of calcium and water-soluble as well as fat-soluble vitamins. Ulcerative colitis does not lead to steatorrhœa, except rarely when there is involvement of the small intestine.

Steatorrhœa can also occur secondary to pancreatic enzyme deficiency, to biliary deficiency or in the so-called malabsorption syndrome. Acute pancreatitis is essentially a surgical emergency; in sub-acute and chronic pancreatitis steatorrhœa may be a presenting syndrome, but in spite of the fact that pancreatic lipases are considered necessary for splitting, the fæcal fat is quite well hydrolysed in this condition. One particular form of this steatorrhœa is extremely important, namely congenital fibrocystic disease. An early diagnosis is essential for effective treatment. The condition is suspected on clinical grounds, the stools show an excess of fat, trypsin cannot usually be detected in the stool and duodenal intubation reveals that there is no amylase or trypsin. It would seem that in this condition there is a general deficiency of all apocrine secretions, including sweat. This has been made the basis of a simple test whereby sweat is collected on filter-paper from a child, the sodium or chloride content is estimated and a figure above 70 mEq/l. for both is said to provide strong support for the diagnosis.

In the steatorrhœa of biliary deficiency splitting again is virtually normal, and it is remarkable that in this condition, although the stools are characteristically fatty in appearance, they are often reasonably well formed. The diagnosis depends upon the presence of jaundice or, when there is cirrhosis of the liver, on the results of liver function tests.

The malabsorption or sprue syndrome comprises sprue cœliac disease of infancy, and non-tropical sprue, (formerly idiopathic steatorrhœa). It was at one time thought that hydrolysed fat—i.e., fatty acids—appeared in the stools in the malabsorption syndrome because of a gross failure of absorption of fatty acids, and since absorption of unhydrolysed fat was then considered not to occur, such patients were believed to be unable to absorb much of their dietary fat. However, loss of weight occurs very late in the malabsorption syndrome, and balance experiments in which the quantity of fat lost in the fæces over a period of days when the patient is on a constant measured intake of fat have revealed that patients with the syndrome absorb between 40 and 85 per cent of the dietary fat, in contrast to normal people, who absorb over 90 per cent. Such findings reveal the fallacies involved in the older determinations of fat in dried fæces, for patients with highly abnormal analysis of dried fæces may show relatively slight impairment of absorption. On the other hand, patients with relatively normal analysis of dried fæces may reveal very severe impairment of fat absorption.

Frazer has shown that in the malabsorption syndrome there is no increase in systemic blood fat after a meal containing neutral fat. It therefore seems that there is a failure in the absorption of particulate fat and that a larger proportion than normal needs to be absorbed as hydrolysed fat. This is, of course, in complete contrast to the earlier views. A curious feature of the malabsorption syndrome is that the fat absorption is proportional to the amount of dietary fat, so that the percentage absorption is independent of the quantity of fat in the diet.

Frazer has found that unsaturated fatty acids are more easily absorbed than saturated fatty acids, and that the percentage absorption of fat is dependent more on the proportion of unsaturated to saturated fats in the diet rather than on the total amount of dietary fat. The fæcal fat thus consists mainly of saturated fatty acids, which form insoluble calcium soaps and may lead to osteomalacia.

It has been known for a long time that in the malabsorption syndrome barium meals lead to a characteristic radiographic deficiency pattern in which the barium sulphate suspension becomes aggregated into segments. It has also been known that excess of long-chain fatty acids in the gut stimulates the secretion of mucus. Frazer has shown that the deficiency pattern is due to the flocculation of the barium sulphate by excess mucin and has obtained similar aggregates of barium sulphate both *in vitro* and in the isolated gut.

Vitamin deficiencies, particularly of thiamine, nicotinic acid, folic acid, riboflavin and pyridoxine, frequently occur in the malabsorption syndrome. These deficiencies come and go, and treatment of one may cause exacerbation of another. Duodenal intubation reveals that whereas in normal individuals very few bacteria are present in the duodenal contents, in the malabsorption syndrome there are large numbers. It has been tentatively suggested that the presence of long-chain fatty acids in the gut results in hypochlorhydria or achlorhydria. The small intestine becomes infected throughout its length by micro-organisms which compete with the host for the dietary vitamins. These bacteria ferment starch and carbohydrate foodstuffs to give short-chain fatty acids which exaggerate the diarrhœa and appear to set up a vicious circle. This is particularly the case in

THE GASTRO-INTESTINAL TRACT

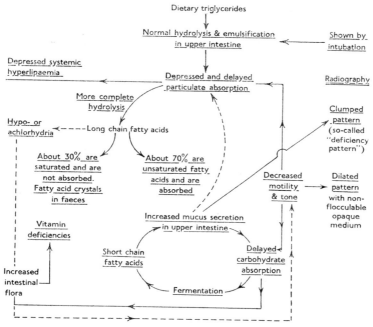

Fig. 18. Malabsorption syndrome (after Frazer).

cœliac disease. The bulky stools of the sprue syndrome may on occasion lead to sodium deficiency, with its attendant dehydration and its sequelæ. Figs. 18 and 19 after Frazer summarize his

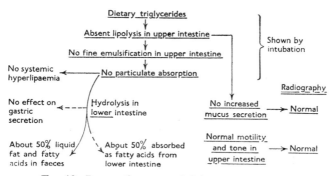

FIG. 19. Pancreatic enzyme deficiency (after Frazer).

views concerning the mechanism of production of the features of the malabsorption syndrome and of pancreatic deficiency respectively.

It is now known that all forms of the malabsorption syndrome show a generalized depression of absorption in the upper small intestine and that there is delay and depression of the absorption curves of glucose, xylose and urea as well as fat. In the xylose tolerance test the subject ingests either 5 or 25 g xylose, and the urinary excretion of xylose in 5 hr. is estimated. The larger dose is unacceptable by some patients. In the malabsorption syndrome less than 1·5 g with the smaller dose and less than 5 g with the larger dose supports the diagnosis of malabsorption syndrome. In all forms of the syndrome there is increased mucus secretion, which causes a characteristic radiological pattern. Cœliac disease is undoubtedly due to a sensitivity to the wheat protein gluten, and children suffering from the disease can be completely relieved of their symptoms when placed on diets free from this protein. A proportion of cases of malabsorption syndrome similarly react favourably to a gluten-free diet, but others do not, and it is not yet known whether these represent a separate disease entity or a persistence of the cœliac syndrome into adult life in which the condition has become irreversible. Although tropical sprue is very similar in its clinical form to the cœliac disease induced by gluten and to the malabsorption syndrome, there is no evidence whatsoever that this condition is in any way related to gluten ingestion. There is a characteristic geographical distribution, and the evidence at present suggests that the condition is due to a change in the intestinal flora.

Laboratory examination in suspected fat absorption defect

Tables 13 and 14 show investigations which may be of value in diagnosis:

TABLE 13—EXAMINATION OF FAT ABSORPTION DEFECT

1. *Fæces*
    (a) Daily fat balance
    (b) Characteristics (colour, bulky, frothy, acid)
2. *Examination of upper intestinal function*
    (a) Intra-luminar condition by intubation
    (b) Rate of absorption, chylomicron counts, glucose tolerance (after intra-duodenal glucose)
    (c) Radiographic appearance: (i) flocculant medium; (ii) non-flocculant medium

TABLE 14

	Pancreatic deficiency.	Biliary deficiency.	Malabsorption syndrome.
Intubation			
Lipase	Absent	Normal	Normal
Bile	Normal	Deficient	Normal
Emulsification	Faulty	Faulty	Normal
Absorption			
Fat (chylomicrons)	Flat	Flat	Flat
Glucose	Normal	Normal	Deficient
Radiology	Normal	Normal	Deficiency pattern

Occult blood tests. When there is massive hæmorrhage into the stomach hæmatemesis usually occurs, with its characteristic coffee-ground vomit, but lesser degrees of hæmorrhage, as well as bleeding from the gut, generally result in the appearance of blood breakdown products in the fæces. Bleeding from hæmorrhoids usually leads to the appearance of bright red blood in the fæces, but this is not the case with hæmorrhage from sites higher up in the alimentary canal. The degradation products of blood may be recognizable as the dark tarry stools of frank melæna or may be required to be identified by special tests for occult blood.

The chemical tests for occult blood can be very sensitive, but detect only those breakdown products of hæmoglobin retaining the iron atom within its molecular structure. On occasions, particularly with hæmorrhage from the upper part of the gastro-intestinal tract, the hæmoglobin may be so broken down that no iron-containing derivatives but only protoporphyrin, deuteroporphyrin or mesoporphyrin are present. These require to be detected by spectroscopic tests, which are also capable of detecting the iron-containing breakdown products, but are less sensitive in this respect than most of the chemical tests. For this reason occult blood tests should include if possible the spectroscopic examination, which is probably sufficiently sensitive for all ordinary work, for the chemical tests are usually so sensitive that they readily detect minor degrees of hæmorrhage from the gums as well as hæm derivatives from the meat of the diet. However, many clinicians prefer the chemical test, which should nevertheless be supported by spectroscopic observations.

The chemical tests all depend on the ability of iron-containing proteins and their breakdown products to catalyse the oxidation of some substance such as benzidine or guaiacum to a blue- or green-coloured pigment. The tests should be performed on specimens of fæces collected on three consecutive days. The benzidine test has hitherto been regarded as the most reliable, but the carcinogenic properties of this reagent has necessitated the use of other reagents including guaiacum and o-tolidine (see page 214). With some the test is performed on a suspension of fæces which has been boiled to destroy peroxidase derived from the vegetable matter of the diet. A purge should be given at the same time that the patient is put on a meat-free and green-vegetable-free diet. The meat-free diet must be free of all red meat or meat extracts, although a little white meat, such as rabbit or chicken, may be allowed. Green vegetables should be excluded, for the appearance of chlorophyll degradation products in the spectroscopic test may confuse those workers without special experience; material derived from green vegetables appears to depress the sensitivity of the chemical tests. Some iron preparations, especially ferrous fumarate, give false positive results.

"Hematest" and "Occultest" reagent tablets contain orthotolidine and a hydrogen-peroxide-generating system. Hematest is less sensitive and is recommended for use with fæces; Occultest is recommended for the detection of blood in urine, but may be used with fæces if the patient has been suitably dieted for a few days before specimens are collected. The "Hemastix" reagent strip is a cellulose strip, one end of which is impregnated with an organic peroxide and orthotolidine.

The main defect of chemical tests for occult blood may be due to the failure to differentiate between the peroxidase activity of blood and peroxidases of dietary origin. Interpretation of results may also be complicated by the possibility that fæces may contain inhibitors and perhaps accelerators of the peroxidase reaction. These difficulties are eliminated by the use of an isotope method for the determination of blood in fæces. In this method the red cells are separated from a sample of blood, labelled with ^{51}Cr and, after washing, reinjected into the circulation. The amount of blood lost into the gastro-intestinal tract can be determined from measurements of radioactivity in the blood and fæces.

A comparative study has shown that the more sensitive tests for occult blood in fæces, including an orthotolidine test, the Hemastix test read at 30 seconds, and the Occultest tablet test give an unacceptably high incidence of positive reactions with the stools of normal subjects taking an unrestricted mixed diet as well as with those of patients losing less than 2 ml. of blood daily, as shown by the radio-chromium-labelled red-cell technique. When these chemical tests are used it is essential to prepare the patient for a few days with a meat- and green-vegetable-free diet, but even then false positives will still occur. The less-sensitive tests, such as the Hematest tablet test and the Hemastix test read at 15 seconds, give fewer positive results with the stools of normal persons not subject to dietary restriction, but comparison with the results obtained with the isotope technique shows that they are not sensitive enough to detect clinically significant fæcal blood losses with any certainty.

If the results of chemical tests for occult blood were regarded as clinical signs to be interpreted along with all the other clinical evidence which in disease can be variable, inconstant and liable to biological variation, then two tests might be used, one of low sensitivity to detect significant amounts of blood in stools, and the other of high sensitivity to distinguish stools containing negligible amounts of blood. Rather than use two different tests, Hemastix might be read at 15 seconds and 30 seconds. However, there is a one-in-five chance that the result may be misleading, and it is questionable whether the chemical tests for occult blood should be retained in modern medicine.

Carcinoid tumours of the intestine. Carcinoid tumours may occur anywhere in the gut or elsewhere but most commonly in the appendix or ileum. They arise from argentaffin cells producing serotonin (5-hydroxytryptamine) and its precursor 5-hydroxytryptophan. Serotonin causes diarrhœa and a characteristic cutaneous flushing and is frequently associated with an increased excretion of 5-hydroxyindolyl acetic acid (5 HIAA).

Further Reading

Avery-Jones, F. 1952. "Modern Trends in Gastro-Enterology," Chapter XII, Part III. Assessment of Gastric Secretion, J. M. Hunt, p. 296. Butterworth, London.

Frazer, A. C. 1962. "The Malabsorption Syndrome, with Special Reference to the Effects of Wheat Gluten." Advances in Clinical Chemistry. Eds. H. Sobotka and C. P. Stewart. Vol. 5. Academic Press. p. 69.

Lubran, M. 1961. "The Augmented Histamine Gastric Function Tests." Association of Clinical Pathologists Broadsheet No. 38 (New Series).

CHAPTER IX

SOME ASPECTS OF THE CHEMICAL PATHOLOGY OF DIABETES

The control of the blood-glucose concentration. In a normal person the concentration of glucose in the blood does not fall much below 70 mg/100 ml. even after many days of fasting, nor rise above about 170 mg/100 ml. after ingesting as much as 500 g carbohydrate in a day. This relative constancy of the blood sugar concentration is made possible by the precisely adjusted balance of glucose entering the blood and glucose leaving the blood (Fig. 20).

Glucose entering the blood is derived from:

(a) dietary carbohydrate;
(b) glycogenolysis of liver glycogen;
(c) gluconeogenesis from protein.

Dietary carbohydrate is absorbed from the intestine mainly as glucose, lævulose or galactose, the last two sugars being readily converted by the liver into glucose. The absorption of glucose from the intestine is controlled by a specific enzymic mechanism and is practically independent of the amount of glucose in the intestine. However, the rate of removal of glucose by the tissues often parallels the rate of intestinal absorption; for example, the increased carbohydrate utilization by the tissues in hyperthyroidism is accompanied by a correspondingly increased rate of absorption of glucose. A proportion of the absorbed glucose is taken up by the liver and stored as glycogen. Glucose passing into the hepatic venous blood may thus be glucose absorbed from the intestine which has not been cleared by the liver from the portal venous blood or, especially under fasting conditions, it may have been secreted by the liver having been formed by glycogenolysis from liver glycogen or by gluconeogenesis from amino acid residues. The liver glycogen is readily converted into glucose by glycogenolysis, i.e., breakdown of glycogen by phosphorylase to glucose-1-phosphate followed by the action of phosphoglucomutase and of glucose-6-phosphatase. Phosphorylase is activated by adrenalin and perhaps by glucagon secreted by the α-cells of the islets of Langerhans.

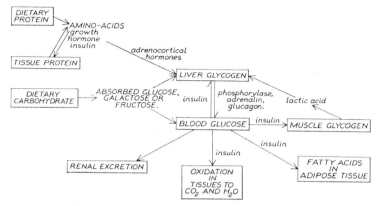

FIG. 20. Control of blood-glucose concentration.

Glucose is removed from the blood by:

(*a*) oxidation to carbon dioxide and water in tissues;
(*b*) formation of fatty acids in adipose tissue;
(*c*) conversion in muscle to glycogen available for degradation to lactic acid;
(*d*) renal excretion if blood glucose concentration is high.

Glucose may be oxidized in the tissue via the glycolytic pathway or the pentose shunt and the Krebs' cycle with the production of carbon dioxide and water and liberation of energy which is utilized for the formation of adenosine triphosphate, the main form in which energy is stored. A large proportion of absorbed glucose is converted in the adipose tissues into fatty acids via acetylcoenzyme A, and a pathway which is virtually the reverse of fatty acid breakdown. A proportion of the blood glucose is converted into glycogen in the muscles, which, however, lack glucose-6-phosphatase so that muscle glycogen can only be converted to glucose by breakdown to lactic acid, part of which can pass in the blood to be reconverted in the liver to glycogen. In the normal individual the glucose presented in the glomerular filtrate is completely reabsorbed, but if the blood sugar is high, so that the amount of glucose passing into the glomerular filtrate exceeds the maximum tubular reabsorptive capacity, or if tubular reabsorption is lowered, either physiologically, as in patients with a low renal threshold, or by phloridzin, urinary loss of glucose may become significant.

The height and direction of changes of the blood glucose are dependent on variations in the rates of entry and removal of glucose along these pathways, each of which is affected by one or more hormones.

Hormonal control of blood glucose. Insulin is the only hormone capable of preventing excessive rise in blood sugar, for it increases carbohydrate utilization by all the metabolic pathways, and may also diminish hepatic output of glucose. The precise mechanisms whereby insulin exerts its profound effect on metabolism are still important research topics. The earlier view that insulin antagonized anterior pituitary hormones in their effects on the hexokinase reaction is no longer accepted. At present it appears that insulin primarily increases cell permeability to glucose and that all its other metabolic effects are secondary to this. However, effects at other metabolic sites cannot finally be excluded, and it seems likely that insulin can increase incorporation of amino acids into protein.

In contrast, the growth hormone of the anterior pituitary gland, 11-oxygenated adrenal steroids and indirectly the adrenocorticotrophic hormone and adrenalin all tend to prevent excessive fall in blood sugar. Growth hormone not only causes a retention of nitrogen and increased protein synthesis but also antagonizes insulin at many of the sites of action. It depresses the glucose uptake of rat diaphragm and prevents the severe loss of muscle glycogen which occurs in fasting hypophysectomized animals. The hypoglycæmic effect sometimes observed with growth hormone in short-term experiments is probably mediated by stimulation of insulin secretion. ACTH stimulates the excretion of hormones from the adrenal cortex, including the 11-oxygenated steroids. These steroids increase gluconeogenesis from the tissue proteins and decrease glucose utilization. Adrenalin prevents a serious fall of blood sugar, since as soon as the concentration falls, the resulting discharge of adrenalin mobilizes liver glycogen by increased glycogenolysis. Adrenalin also stimulates the secretion of ACTH, and therefore of adrenal steroids, which in turn accelerate the conversion of tissue protein into liver glycogen.

Hypoglycæmia. From these considerations it is clear that hypoglycæmia will occur if there is either excessive insulin secretion, which may or may not be due to a tumour of the islets of Langerhans, or deficiency of those hormones concerned with maintaining blood sugar. Thus, the blood sugar readily falls to low levels after a short period of fasting in patients with Addison's disease or Simmonds' disease, in which there is deficiency of adrenal cortical secretion or of anterior pituitary hormones respectively.

Hyperglycæmia. Conversely, hyperglycæmia will occur if there is:

(a) absolute or relative deficiency of insulin secretion;
(b) excessive secretion or administration of adrenal cortical steroids;
(c) excessive secretion of ACTH, as in Cushing's syndrome;
(d) excessive secretion of growth hormone during the active phases of acromegaly and gigantism;
(e) secretion of adrenalin and *nor*-adrenalin by tumours of the adrenal medulla;
(f) hyperthyroidism and simple fasting, which cause the blood sugar to rise to excessively high levels after administration of carbohydrate by mouth.

Insulin in plasma. The very small amounts of insulin in plasma may be detectable by their effect on the uptake of glucose by rat diaphragm, the effect on the synthesis of radioactively labelled fat from labelled glucose in adipose tissue or more recently by immunochemical methods (see pp. 83 and 84). Other methods have used diabetic animals rendered highly sensitive to insulin by adrenalectomy and/or hypophysectomy. Plasma contains a number of inhibitors of insulin, e.g. the synalbumin of Vallance-Owen, and some of these are present in considerable amounts in diabetic ketosis, during infections and in association with severe insulin resistance due to insulin antibodies.

Causes of diabetes mellitus. Many patients with diabetes mellitus require much more insulin than do patients made diabetic by total pancreatectomy for cancer. Moreover, insulin-like activity has been shown to be present in the blood plasma of many diabetic patients, but not usually in that of young ketosed diabetic subjects. Diabetes may therefore be due either to an absolute lack of insulin (due to failure of the β-cells of the islets of Langerhans) or to a relative lack, brought about by an extra-pancreatic factor such as an excess of antagonizing hormones or other insulin-inhibiting factors.

The diabetogenic activity of growth hormone in dogs and cats is firmly established, but the only evidence of its role in human diabetes is provided very indirectly by the high incidence of diabetes in women who have produced overweight babies. The overweight babies and the subsequent development of diabetes might be due to growth hormone, but there is no direct evidence of this. Excessive growth hormone production can be associated with diabetes during the active stages of acromegaly and gigantism, and a diabetic state can be induced or exaggerated in Cushing's disease or during adrenal cortical hormone or ACTH therapy. Panhypopituitarism or adrenal insufficiency can modify, but does not cure, diabetes in man, nor do adrenalectomy or hypophysectomy, which have been performed in an attempt to alleviate the vascular complications. There is indeed no direct evidence that either the pituitary or the adrenal glands play a causal role in human diabetes mellitus.

The presence of insulin in the plasma of elderly obese diabetics, and in some young ketosed diabetic subjects has emphasized the role of insulin antagonists as a cause of diabetes. Such antagonists presumably cause the islet tissue to over-secrete insulin and eventually fail. Recent developments suggest that in addition to the insulin antagonists referred to above, the presence of non-esterified fatty acids in plasma can interfere with the action of insulin and cause excessive secretion of insulin and finally failure of insulin secretion.

Elderly, obese diabetics often regain normal glucose tolerance after weight reduction, and this is understandable according to a recent concept that there is a reciprocal relationship between glucose and fatty acid metabolism. The depression of carbohydrate metabolism is believed to be secondary to the excessive release and oxidation of fatty acids in these subjects.

There is no doubt that diabetes mellitus is no longer to be regarded as a disease but as a syndrome which can result from a number of disease processes. In the near future the estimation of insulin, of growth hormone and of other insulin antagonists by sensitive methods is likely to change greatly our concepts of diabetes.

Metabolism in diabetes. The main disturbances of metabolism in diabetes are:

1. An absolute or relative insufficiency of insulin leading to deficient utilization of carbohydrate and excessive glycogenolysis and gluconeogenesis, causing hyperglycæmia and glycosuria.
2. The decreased utilization of glucose for the production of energy by the tissues increases the utilization of fat, which is hydrolysed in the adipose tissues with the formation of glycerol (which can be converted into carbohydrate) and fatty acids. These are released as non-esterified fatty acids (NEFA) into the bloodstream, where they are taken to the other tissues for utilization, or to the liver for esterification and liberated again into the bloodstream as fats.
3. The fatty acids are degraded in the liver into acetyl-coenzyme A units, which if sufficient oxaloacetic acid and reduced nicotinamide adenine dinucleotide phosphate (NADPH*) are available from carbohydrate oxidation can be oxidized in the Krebs cycle. If not, acetylcoenzyme A may be converted into aceto-acetylcoenzyme A or be synthesized into cholesterol. Aceto-acetylcoenzyme A may be converted into acetoacetic acid, which with the reduction product β-hydroxybutyric acid and its decarboxylation product acetone form the so-called ketone bodies. Aceto-acetic acid can be utilized by the tissues, especially the muscles, for the production of energy, but if it is formed by the liver at a rate greater than the muscles can utilize it, then ketone bodies accumulate in increased quantities in the blood and ketosis develops.
4. The increased concentration of non-esterified fatty acids in

---

* NADPH (TPNH or coenzyme II as it was formerly called) is also necessary for the synthesis of fatty acids and is provided by the breakdown of glycogen via glucose-6-phosphate and the pentose shunt. NADPH is necessary for the incorporation of pyruvate in the Krebs cycle, the re-synthesis of carbohydrate from pyruvate, gluconeogenesis from amino acids and for the biosynthesis of cholesterol and steroids.

the plasma (which is a primary event in obese diabetics) causes impaired sensitivity to insulin, impaired pyruvate tolerance, excessive conversion of glucose to muscle glycogen rather than to pyruvate, and therefore exaggerates the impairment of glucose tolerance.

Glycosuria. Reducing substances found in urine which may be confused with glucose include:

(*a*) lactose, pentoses and lævulose;
(*b*) creatinine and uric acid in concentrated urines;
(*c*) substances excreted in conjugation with glucuronic acid, e.g., acetanilide, antipyrin, camphor, chloroform, chloral, morphine, menthol, phenol, pyramidon and turpentine;
(*d*) salicyluric acid and gentisic acid, which have very weak reducing properties and are excreted after salicylate therapy;
(*e*) homogentisic acid.

In most cases it is possible to get some idea from the clinical history as to the likelihood of the reducing substances being due to the ingestion of drugs. Homogentisic acid is excreted in the extremely rare condition alkaptonuria, characterized by normal health, except for the urine turning black on standing. Homogentisic acid is by far the strongest reducing agent found in urine, and readily reduces alkaline silver nitrate to metallic silver in the cold. The differentiation between glycosuria, lactosuria, pentosuria and lævulosuria requires special consideration before doing elaborate laboratory tests. It is important to test the urine for ketone bodies either by Acetest or Rothera's test for acetone and aceto-acetic acid and by the ferric chloride test for aceto-acetic acid, taking care that the colour given by salicylates is not confused with a positive reaction. Apart from the fact that salicylates give a colour with ferric chloride even after boiling, it is useful to remember that a strong ferric chloride reaction due to aceto-acetic acid will be accompanied by a strongly positive Acetest or Rothera reaction, and if this is not the case, salicylates should be suspected. In the presence of severe ketosis, there is usually no point in carrying out any laboratory tests, for the typical clinical history of loss of weight, pruritus, thirst and polyuria will substantiate the diagnosis of diabetes, and the severe ketosis is an indication for beginning insulin treatment immediately after a single determination of the blood sugar to confirm the diagnosis. This single blood sugar determination will be well above the normal range. If reducing substances are present in the urine in the absence of ketosis a single blood sugar determination above the normal range is often sufficient to confirm the diagnosis if the clinical history is typical. There is then no point in doing laboratory tests either to confirm that the reducing substance is glucose or to confirm that diabetes is present.

In a small percentage of cases a typical history may not be obtained and slight or no ketosis is present; single determinations of blood sugar are then often unhelpful, for they usually give border-line results difficult to interpret. In such circumstances the reducing substance should be shown to be glucose by paper chromatography or an enzyme test (see page 209). Indeed, the development of simple and specific enzyme tests for glucose (see appendix) has profoundly altered the problem of assessing the significance of reducing substances in urine. Identification by fermentation or by osazone formation is unsatisfactory. Pentoses may be examined by Bial's test and lævulose by Seliwanof's test. If the glycosuria is confirmed it may then be necessary to do a glucose tolerance test, but this is only essential when an atypical history is present, and is usually of greater importance in the exclusion rather than in the confirmation of diabetes. Because of the impairment of glucose tolerance occurring during fasting—i.e., because of the excess gluconeogenesis occurring in such circumstances—it is necessary to give the patient a full carbohydrate diet for 7 days before test. After fasting overnight, blood and urine specimens are analysed quantitatively and qualitatively for glucose before, $\frac{1}{2}$, 1, $1\frac{1}{2}$ and 2 hours after 50 g of glucose in 200 ml. water.

Persons with a normal glucose tolerance have a fasting blood sugar below 120 mg/100 ml., which does not rise above 170 mg/100 ml. and returns to normal within 2 hours. Glycosuria is absent throughout. A second type of normal curve is the so-called "lag storage curve", in which the $\frac{1}{2}$-hour blood analysis shows an abnormally high concentration of glucose, but the 1-hour specimen has returned to normal. Glucose may be present in the urine in the $\frac{1}{2}$- and 1-hour specimens (see Fig. 21). Some patients have an abnormally low reabsorption of glucose which appears in the urine at times when glucose is not abnor-

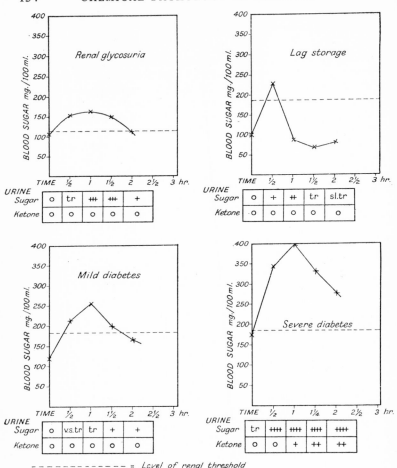

Fig. 21. Glucose tolerance curves.

mally high. These are the so-called renal threshold cases. Impaired glucose tolerance is present when the fasting blood sugar is above 130 mg/100 ml., the blood sugar rises to a level above 170 mg/100 ml. and the return to normal values is delayed to 3 hours or more. Impaired glucose tolerance is observed not only in diabetes mellitus but also in hyperthyroidism, severe liver disease, Cushing's syndrome, some cases of acromegaly, sepsis and carbohydrate starvation.

The chemical pathology of diabetic coma. Diabetic coma may arise as the first recognized sign of diabetes of acute onset in a hitherto unrecognized diabetic, or, more usually, as a result of some infection or other precipitating cause in a known diabetic under treatment. In either event there results a well-defined sequence of changes in composition of the various fluids of the body which require to be fully appreciated for the formulation of a rational line of treatment. The sequence of events leading to and persisting during diabetic coma is:

(a) hyperglycæmia and glycosuria secondary to deficient carbohydrate utilization because of absolute or relative lack of insulin;

(b) hyperlipæmia due to increased non-esterified fatty acids, glycerides and cholesterol in the plasma;

(c) ketonæmia and ketonuria because of increased production of ketone bodies by the liver at a greater rate than the muscles can use them;

(d) acidosis and acidæmia;

(e) polyuria due to the large quantities of glucose excreted in the urine, with consequent loss of water by the body and development of dehydration and hæmo-concentration; the consequent loss of water is borne by both extracellular and intracellular compartments of the body fluids;

(f) increased urinary loss of nitrogen, phosphate and potassium.

(g) a fall in extracellular fluid volume due to sodium deficiency associated with a fall of plasma volume, a fall in blood pressure, peripheral circulatory failure, renal dysfunction and retention of end products of metabolism.

Cause of diabetic coma. Diabetic coma is not produced by the raised concentration of glucose in the blood, since lower levels of blood sugar may sometimes be found in coma than in non-comatose patients. On the other hand, in coma high concentrations of ketone bodies are always present in the blood, and, except when there is very severe renal dysfunction, in the urine. A few cases without ketonuria or ketonæmia have been described, but in these either the presence of ketone bodies in the blood had not been investigated by a reliable test or the diagnosis was in doubt. It is unlikely that the acidosis itself is the cause of coma, since acidosis of corresponding severity can be produced by means other than ketosis and yet be unaccompanied by coma of the type found in diabetic ketosis. Aceto-acetic acid is the first formed ketone body and the most important because it can exist in an enolic form. Acetone and β–hydroxybutyric acid are said to be non-toxic and are formed secondarily by decarboxylation and reduction respectively of aceto-acetic acid. Diabetic coma is therefore brought about either by a direct effect of aceto-acetic acid on the central nervous system or by some unknown disturbance of the dynamic equilibria of the body associated with the production or disposal of ketone bodies.

Concentration of glucose in the blood. Although providing invaluable information on the response of the patient to therapy, this does not provide an absolute means of control of treatment, as might be thought. The amount of insulin required to control the blood sugar is dependent upon many factors, including the previous insulin requirement of the patient, the height of the blood sugar, the presence of infection and insulin sensitivity or resistance. A substantial dose of insulin is therefore administered at regular intervals until the patient has recovered consciousness. Measurements of the blood glucose during the early stages of treatment will indicate the response of the blood sugar to the initial doses of insulin and indicate the need for adjustment of the size of the subsequent doses. Repeated measurements of blood glucose are very helpful in following treatment because the body reserves of carbohydrate may be used up before ketonæmia has been abolished. The patient may then pass from a state of diabetic coma to one of hypoglycæmic coma, sometimes without even recovering consciousness. Some workers administer glucose with each dose of insulin. The need for repeated blood sugar estimations then becomes less imperative during the early treatment of diabetic coma, although in the later stages, when recovery is beginning and ketonuria and ketonæmia disappear, repeated estimations are necessary in order to determine how soon the frequent injections of glucose and insulin may be reduced. Not all workers are in agreement with the wisdom of administration of glucose with insulin.

Concentration of glucose in the urine. This normally depends on the concentration of glucose in the plasma, the glomerular filtration rate, the maximum renal tubular reabsorptive capacity for glucose (glucose T_m) and the factors controlling the urinary volume, such as hydration, dehydration, salt deficiency and tubular reabsorption of water. In diabetic coma, when the concentration of glucose in the plasma is very high, the concentration of glucose in the urine will depend on the maximum osmotic work of which the kidneys are capable, as well as on other substances requiring excretion. It is obvious, therefore, that with blood sugars considerably above the renal threshold for glucose, the concentration of glucose in the urine will bear no direct relationship to the concentration in the blood, and in some cases of diabetic coma with renal impairment due to salt deficiency and secondary dehydration, no glucose whatever may be present in the urine. The accurate estimation of the concentration of glucose in the urine is therefore unnecessary in the laboratory control of the treatment of diabetic coma; sufficient information is obtained from simple qualitative tests for glucose in urine, especially when the results are interpreted in a semi-quantitative manner (i.e., +, ++, +++ and ++++).

Concentration of ketone bodies in the blood and urine. The ketone bodies in the urine bear no constant relationship to the concentration of ketone bodies in the blood because the three components of ketone bodies are excreted differently. Acetone is a non-threshold substance, but β-hydroxybutyric and aceto-acetic acids are excreted as threshold substances. The proportions of the three ketone bodies in the blood differ considerably from one patient to another, and must depend on the rate of production of aceto-acetic acid, the rates of its conversion into acetone and β-hydroxybutyric acid, the rates of utilization and on the different ways of excretion. The relationship between urinary ketone bodies and blood ketone bodies is therefore a very complicated one. The determination of the proportions of the three ketone bodies is not easy, especially in an emergency, and it is fortunate that the results of qualitative urine tests for acetone and aceto-acetic acid usually roughly parallel the total urinary excretion of ketone bodies and, except when there is severe impairment of renal function, provide an adequate guide to the concentration in the blood.

Acidosis and acidæmia. The plasma bicarbonate concentration bears no direct relationship to the blood ketone content, because, of the three ketone bodies, only one, aceto-acetic acid, is strongly acidic. The depression of plasma bicarbonate affords the most accurate index of the severity of the acidosis. The plasma bicarbonate, however, will also be affected by the azotæmia resulting from sodium deficiency (see below). In the initial stages of diabetic ketosis the acidosis is compensated by a fall in P_{CO_2}, and the pH of the body fluids remains within the normal range. In later stages, however, the fall in plasma bicarbonate becomes so extreme that the P_{CO_2} fails to decrease proportionately; the pH of the body fluids then falls below the lower limit of normal As the acidosis and acidæmia of diabetic ketosis are non-respiratory in origin, the state of acid–base balance of the body fluids may be adequately followed by measurements of the plasma bicarbonate.

Blood lipids. The increased blood lipids are of considerable theoretical interest, but their measurements are of no value in the diagnosis and treatment of diabetic coma. However, the increase in plasma cholesterol may sometimes be sufficient to cause xanthomatosis.

Plasma protein and hæmoglobin. The polyuria of uncontrolled diabetes may lead to simple dehydration, in which the water loss is shared by the entire body-water. (This is in contrast to the dehydration secondary to salt deficiency, when the water loss is borne mainly by the extracellular fluid.) This simple dehydration may result in raised plasma protein, blood hæmoglobin concentrations and hæmatocrit readings.

Serum sodium. The anions of the ketone acids are excreted in the urine, and would be accompanied by an equivalent quantity of sodium ions if they were not exchanged for ammonium and hydrogen ions in the renal tubules (see page 27). The conservation of the sodium ions is never complete, and there results a decrease in the body content of sodium, which usually leads to a fall of the concentration of sodium in the plasma. This fall may be diminished or even changed to an increase by the dehydration secondary to the sodium loss, as well as by the dehydration produced by the polyuria caused by glycosuria. The actual total amount of sodium present in the extracellular fluid is clearly more important than the concentration of sodium in the plasma.

Plasma chloride. The electrolyte disturbances of the body fluids brought about by ketosis result in a lowered concentration of sodium in the extracellular fluid, but *per se* there is no effect on the concentration of chloride ions. Patients in diabetic coma frequently vomit in the early stages, and the chloride loss in the vomit probably accounts for the fall in plasma chloride frequently found. Occasionally the concentration of chloride ion in the plasma is found to be elevated, and presumably in these cases there has been a minimal loss of chloride by vomiting, and the dehydration caused by the deficiency of sodium and by the glycosuria may produce this hyperchloræmia.

Plasma potassium. In many conditions in which there is loss of extracellular sodium there is also loss of intracellular potassium in the urine, so that the secondary dehydration associated with sodium loss is not borne entirely by the extracellular fluid (see page 52). A transfer of potassium from the cells to the extracellular fluid continues throughout the onset and development of diabetic coma and is accompanied initially by a raised concentration of potassium in the plasma and increased excretion in the urine. With appropriate correction of the hyperglycæmia, ketosis, sodium and water deficiencies, a deficiency of potassium in the body reveals itself by weakness and even paralysis of the skeletal musculature and respiratory difficulty. Analysis reveals that the concentration of potassium in the plasma is now low, the deficiency of the body potassium having been unmasked by the rehydration. The fall in plasma potassium may be exaggerated by the action of insulin, which is known to transfer potassium from the plasma to the cells, thus decreasing the plasma and urinary potassium. Careful balance experiments have shown that in diabetic coma the urine always contains potassium ions, even when there is depletion of potassium in the cells as well as in the plasma. They also show that when potassium salts are administered during the treatment of diabetic coma potassium is retained and passes into the intracellular compartment.

Blood urea. The reduction in magnitude of the extracellular compartment associated with the sodium loss appears to be intimately related to a fall in plasma volume, peripheral circulatory failure with low blood pressure and renal failure with retention of end products of metabolism such as urea.

Plasma phosphate. In diabetic coma there is an increased concentration of inorganic phosphorus in the plasma and an increased excretion of phosphate in the urine. This loss of phosphate in the urine may lead to phosphorus deficiency, which, like potassium deficiency, is unmasked by rehydration during treatment. Insulin treatment alone reduces the plasma concentration to levels below normal. This has been shown not to be due to hypoglycæmia, since it is not inhibited by the simultaneous administration of glucose. However, phosphorus deficiency may impede carbohydrate utilization, and claims have been made that the administration of a phosphate buffer during treatment of diabetic coma greatly reduces mortality. It is difficult to see the logic of this—the body contains vast stores of readily available phosphorus in the skeleton, and carbohydrate utilization is not impaired in hyperparathyroidism in which there is also hypophosphatæmia.

Antidiabetic drugs. A number of oral antidiabetic drugs (e.g., tolbutamide) are effective in controlling the glycosuria of unketosed elderly diabetics (usually but not always obese) and in whom the plasma can be shown to contain insulin-like activity. These antidiabetic drugs appear to enhance insulin activity, but whether this is by a stimulating action on the β-cells, by an indirect action on inhibitors or by delaying the inactivation of circulating insulin by insulinase is still unknown.

The detection of diabetes. Much effort has been put into the early detection of diabetes in an apparently symptomless population. The screening tests used have included the detection of glycosuria, the estimation of the fasting blood sugar followed by confirmation with a glucose tolerance test when necessary. Some workers have claimed that the prediabetic state can be best recognized by the abnormally decreased glucose tolerance after cortisol therapy, by the increased concentration of non-esterified fatty acids or even insulin in the plasma of the fasting prediabetic subject. However, these investigations are best carried out in special departments, as these findings have not yet been definitely confirmed. There is no doubt that unrecognized diabetes is at least as common as the overt form of the disease.

Further Reading

Ciba Foundation Colloquia on Endocrinology. 1964. Volume 15. "The Aetiology of Diabetes Mellitus and its Complications." Churchill, London.

LAWRENCE, R. D. 1955. "The Diabetic Life." Churchill, London.

"Insulin," ed. by F. G. Young, *Brit. Med. Bull.*, 1960, **16**, No. 3.

RANDLE, P. J., GARLAND, P. B., HALES, C. N. and NEWSHOLME, E. A. 1963 "The Glucose Fatty Acid Cycle. Its Role in Insulin Insensitivity and the Metabolic Disturbances of Diabetes Mellitus." *Lancet*, **1**, 785.

'Symposium on Insulin." 1966. *Am. J. Med.* **40**, 651.

CHAPTER X

BIOCHEMICAL TESTS IN ENDOCRINE DISEASE

The Thyroid Gland

THE essential function of the thyroid gland is the synthesis and secretion of the thyroid hormones (Fig. 22). Iodide is removed from the bloodstream and concentrated in the gland. The iodide

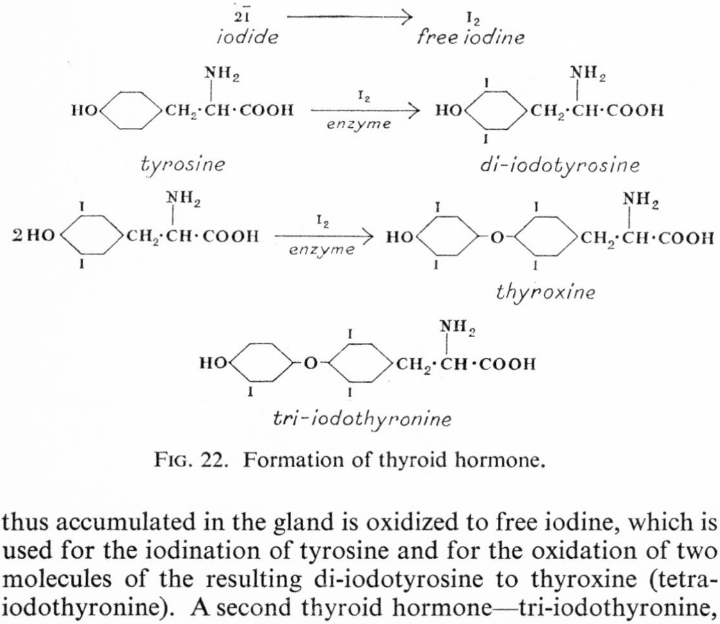

$$2\bar{I} \xrightarrow{} I_2$$
iodide   *free iodine*

tyrosine $\xrightarrow[enzyme]{I_2}$ di-iodotyrosine

2 di-iodotyrosine $\xrightarrow[enzyme]{I_2}$ thyroxine

tri-iodothyronine

FIG. 22. Formation of thyroid hormone.

thus accumulated in the gland is oxidized to free iodine, which is used for the iodination of tyrosine and for the oxidation of two molecules of the resulting di-iodotyrosine to thyroxine (tetra-iodothyronine). A second thyroid hormone—tri-iodothyronine, is formed from one molecule each of di-iodotyrosine and mono-iodotyrosine. The antithyroid drugs, such as thiouracil, commonly used in the medical treatment of hyperthyroidism act by inhibiting the enzyme system responsible for the oxidation of the iodide ion to free iodine; others, such as thiocyanate however, either inhibit the actual concentration of iodide in the thyroid or act on other stages in the synthesis of the thyroid hormone.

The circulating thyroid hormones are bound to protein (thyroxin to a specific thyroxin-binding globulin TBG, a pre-albumin fraction TBPA, as well as to albumin; tri-iodothyronine to TBG and albumin only). Despite this complexity, the determination of the so-called protein-bound iodine of the plasma provides a useful measure of the circulating thyroid hormone.

The release of the hormone by the gland is normally stimulated by the thyrotrophic hormone (TSH, thyroid-stimulating hormone) of the anterior pituitary gland by a feedback mechanism. The secretion of TSH is stimulated when the concentration of the circulating thyroid hormone is low, and is suppressed when there is an excess of circulating thyroid hormone. A second thyroid-stimulating hormone is present in the γ-globulin fraction of the plasma of patients with thyrotoxicosis. Its origin is not known, and since its effects in animals last longer than those due to TSH, it is known as the long-acting thyroid stimulator (LATS). This may be responsible for the exophthalmos as well as the hypersecretion of thyroid hormones in thyrotoxicosis.

The basal metabolic rate. One function of the thyroid hormones is a general stimulation of metabolism, which is most readily assessed in the human subject by determination of the basal metabolic rate. When the subject is fasting and at physical and mental rest the rate of metabolism as measured by the calories of heat produced per unit of time is reasonably constant for any one individual, and when expressed as kilo-calories per hour per square metre of body surface is reasonably constant for normal individuals of the same age and sex.

For technical reasons it is impractical as a routine procedure to determine the heat production of a patient, and indirect means need to be used. One way in which this may be done is to collect for a measured period of time the expired air and determine its volume and carbon dioxide and oxygen contents. From the known composition of the inspired air the volume of the inspired air can be calculated, and thence the oxygen consumption and carbon dioxide production. From the ratio of these—the Respiratory Quotient—and on the assumption that only carbohydrate and fat are being oxidized, the proportions of these two foodstuffs being oxidized can be calculated, and thus the heat production in kilo-calories per hour.

In a simpler and more widely used technique the oxygen consumption is measured in a spirometer. A standard Respiratory Quotient is assumed, and standard tables of oxygen consumption are available for normal individuals of every age, height and weight. The patient's oxygen consumption is expressed as a percentage increase or decrease of the normal and provides a comparative measure of the basal metabolic rate (B.M.R.).

The wide range— $+20$ to -20 per cent in normal persons— is in part due to normal variation, but also because some patients are nervous and find difficulty in adjusting themselves to basal conditions. A value below the normal range usually indicates hypothyroidism; a high value may indicate hyperthyroidism but can also occur in other conditions, such as leukæmia. The measurement of B.M.R. is now little used.

Plasma protein-bound iodine. An estimation of the protein-bound ^{131}I after administration of this isotope was formerly used as a simple diagnostic test, but the plasma protein-bound iodine (PBI) is now readily measured and provides a useful indication of the concentration of circulating thyroid hormones. The normal range is 3–8 $\mu g/100$ ml., but the PBI is affected by a large number of drugs, including not only those containing iodine but also those suppressing ovulation; it is also affected in pregnancy and in a number of general illnesses, including liver and kidney disease and debilitating illness.

Since the PBI includes the iodine of the two thyroid hormones as well as non-hormonal iodine (normally little in amount) and since the binding of the two hormones to plasma proteins is complex (see p. 143), the relationship between the PBI and thyroid status is less direct than desirable. Measurement of free thyroxine and tri-iodothyronine is desirable but technically impractical for the routine laboratory; thyroid-binding capacity of the plasma proteins is more practical and involves measurement of the displacement from the hormone–protein complex by a radioactively labelled hormone. Another method of assessing protein binding depends upon measuring ^{131}I tri-iodothyronine distribution between the plasma and either erythrocytes or an anion exchange resin, both of which contain combining sites and compete for the hormone. However, the detailed distribution of the two thyroid hormones between the free state and the bound state with the three plasma proteins concerned is highly complex.

Radioactive iodine tests. These tests measure aspects of thyroid function different from that indicated by the B.M.R. determination. The latter determination essentially measures the overall response of the tissues to the circulating thyroid hormone, whereas, in contrast, most of the radio-iodine tests measure directly or indirectly the ability of the thyroid to concentrate iodide. Those in common use are as follows:

1. Determination of the accumulation of radio-iodine in the thyroid by direct measurement of radioactivity in the neck region.
2. Determination of the rate of removal of radio-iodine from the blood. In combination with (1) a measure of the thyroid clearance of iodine is obtained (see Chapter I).
3. Determination of the urinary excretion of radio-iodine; the excretion varying inversely to the thyroid uptake.
4. Determination of the rate of release of radio-iodine from the thyroid after further uptake has been blocked by an antithyroid agent.

The significance of ^{131}I uptake tests is enhanced if the plasma inorganic iodide is estimated; this allows the absolute uptake of iodine to be calculated and corrects for changes due to alterations in plasma iodide.

TSH and LATS. These two hormones, the origin of only the first of which is known, may be measured biologically and more recently by immuno-chemical methods. The latter methods are still being developed, but have already thrown light on the feedback control of thyroid secretion. Their clinical importance is not yet known.

Hyperthyroidism. In some cases the thyroid itself may be at fault by responding excessively to a normal secretion of TSH by the pituitary. In others, the condition is only rarely secondary to over-production of TSH but more frequently to the secretion from an unknown origin of LATS which is also probably responsible for exophthalmos.

The B.M.R. is now rarely performed, but measurements of

PBI, radioactive iodine uptake and in difficult cases thyroxine-binding capacity of the plasma proteins are of importance in confirming diagnosis. Measurements of TSH and LATS are available only in special centres.

The average fasting plasma cholesterol concentration of a series of hyperthyroid patients is significantly lower than the average of a series of normal individuals, but the estimation is useless in assisting the diagnosis in any individual patient because of the wide normal range.

Hypothyroidism. Hypothyroidism may occur because of primary thyroid deficiency, as is usually the case in myxœdema and cretinism, or secondary to TSH deficiency, which is almost always one aspect of panhypopituitarism (Simmonds' disease), a condition which will be considered in detail later. Myxœdema is usually diagnosed on clinical grounds (i.e., gain in weight, dryness of the skin, lethargy) and confirmed by the low PBI.

It is possible to distinguish between primary thyroid deficiency from myxœdema of pituitary origin; injection of TSH in the latter condition greatly increases the uptake of radioactive iodine by the thyroid.

If cretinism is to be effectively treated it must be diagnosed during early infancy, when a B.M.R. determination by ordinary methods is impossible and the use of ^{131}I undesirable. Measurement of PBI or even of the fasting plasma cholesterol is thus of importance, the latter being especially valuable in the long-term control of treatment which should be directed to reducing this to the normal range of 120–250 mg/100 ml.

The adrenal cortex

The carbohydrate active or glucocorticoid steroids secreted by the adrenal cortex in man are cortisol and small amounts of corticosterone* (Fig. 23). The main salt-retaining steroid is aldosterone (corticosterone with an aldehyde group in the 18-position), while the main androgens secreted by the adrenal cortex are dehydroepiandrosterone (DHA) and androstenedione (and its 11-hydroxy derivative), all of low biological potency compared with the testis hormone, testosterone (Fig. 25).

* Cortisol and corticosterone differ only in the 17-OH group of the former.

The secretion of cortisol is controlled by adrenocorticotrophin (corticotrophin, ACTH), the secretion of which by the pituitary is influenced by corticotrophin-releasing factors of the hypothalamus; aldosterone secretion is controlled by the renin–

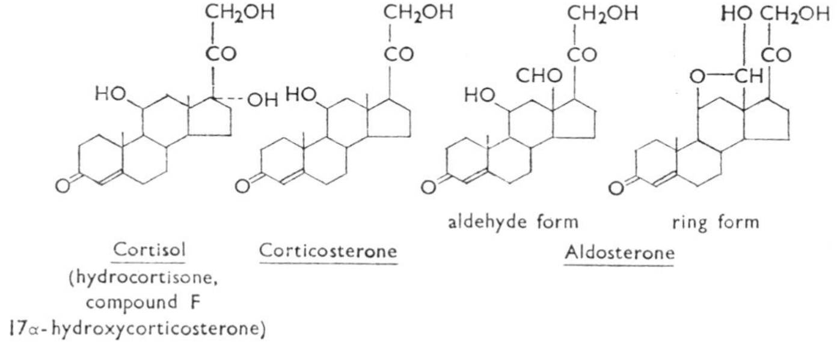

FIG. 23. The naturally occurring adrenal steroids.

angiotensin mechanism, but the control of the adrenal androgens is not known, although it is probably indirectly related to FSH secretion by the pituitary and perhaps to androgen excretion by the gonads.

The metabolism of adrenal cortical hormones. The steroid hormones in general are metabolized to a variety of steroid compounds, some of which are excreted in the free state, while others are excreted in conjunction with glucuronic acid or sulphuric acid.

The glucocorticoid hormones secreted by the adrenal gland are converted into at least thirty other substances by metabolic changes. In urine at least five to ten times as much material is excreted in conjugation with glucuronic acid as is excreted in the free state. Less than 5 per cent of the cortisol secreted by the adrenal is converted to 17-oxosteroids. The 17-oxosteroids derived from the glucocorticoid hormones all bear an 11-OH or 11-oxo group.

Aldosterone and its reduction products are excreted as conjugates. Methods of determination are complicated and difficult, and the routine techniques estimate only free aldosterone and a particularly labile conjugate.

The androgens secreted by the adrenal cortex are either

excreted as the glucuronide or sulphate conjugate (mainly DHA) or metabolized to other compounds in other parts of the body.

Laboratory investigations in adrenal and pituitary disease

The neutral 17-oxosteroids (17-OS) of the urine. These consist of a complex mixture of steroid hormones, all characterized by an oxo group in the 17-position. Although the total excretion of neutral 17-oxosteroids may easily be measured, their complexity, their origin from both adrenals and testes, and technical difficulties in hydrolysing their conjugates makes the estimation of value only when there is gross departure from the normal, as in the diagnosis and treatment of the adrenogenital syndrome and in the diagnosis of adrenal tumours. The normal urinary excretion of 17-oxosteroids depends upon age, but varies from 7 to 27 mg/day and from 4 to 19 mg/day in adult males and females respectively. The excretion is lower in childhood. Determinations of the various fractions of the urinary 17-oxosteroids are time-consuming and may be performed only by departments with special facilities.

The 17-hydroxycorticosteroids of urine (17-OHS, total 17-oxogenic steroids). The majority of the urinary metabolites of cortisol may be estimated as a group by oxidation to 17-oxosteroids, in which form they are measured. These metabolites are therefore called 17-oxogenic steroids. A modification of this technique estimates all adrenal steroids containing a hydroxyl group in the 17-position (17-hydroxycorticosteroids or *total* 17-oxogenic steroids); these include a group of steroids not determined in the original method. In normal subjects without disturbances of adrenal biosynthesis these additional steroids are negligible, and the 17-oxogenic steroid and 17-hydroxycorticosteroid analyses give similar results. These estimations are particularly valuable when carried out before and after stimulation or suppression of the adrenal cortex or pituitary gland. The normal excretion of 17-hydroxycorticosteroids is age dependent and amounts to 10–20 mg/24 hr. in adult males and 5–13 mg/24 hr. in adult females. The 17-oxogenic steroids are usually about 2 mg/day less.

Plasma cortisol. This is now readily measurable in a number of laboratories. The normal range is from 7 to 25 μg/100 ml.,

BIOCHEMICAL TESTS IN ENDOCRINE DISEASE 149

and there is a diurnal variation with a minimum during the three to four hours after midnight and a maximum around 08.00 hours in the morning. However, plasma estimations reflect adrenal activity only at the precise time that the blood is collected, and the diurnal variation and lability of the plasma cortisol in response to psychological and physical stress can make interpretation of results of single determinations difficult.

Adrenal stimulation by ACTH.* Plasma cortisol and urinary 17 OHCS normally increase in response to the injection of ACTH or the synthetic polypeptide, Synacthen; the latter is safer and acts more quickly than ACTH.

Adrenal suppression by dexamethasone or betamethasone.* These strongly active synthetic glucocorticoids in doses too small to interfere with urinary or plasma analyses operate the negative feedback control of cortisol secretion by suppressing ACTH secretion by the pituitary, and therefore cortisol secretion by the adrenal.

The effect of metopirone.* This compound inhibits 11-hydroxylation, the final step in cortisol synthesis; the gland therefore secretes the mineralocorticoid compound S (cortisol without the 11-hydroxyl group). This is inactive in the feedback mechanism and does not suppress ACTH secretion. Compound S secretion by the adrenal therefore increases and leads to an increased urinary excretion of its metabolites, which are measurable as 17 OHCS (compound S is, however, not measured by the fluorimetric methods for the estimation of plasma cortisol).

Stimulation of the cerebro-hypothalamo-pituitary-adrenal pathway by insulin.* Insulin-induced hypoglycæmia normally stimulates the hypothalamus or higher centres to cause an increase of ACTH secretion and therefore of cortisol secretion. This is believed to be a test of the integrity of cerebro-hypothalamo-pituitary pathway.

Lysine–vasopressin test. The synthetic polypeptide—lysine vasopressin—acts like a corticotrophin-releasing factor, and its effect on ACTH and cortisol secretion is used to assess the ability of the pituitary to release ACTH.

* For details see appendix.

150 BIOCHEMICAL TESTS IN ENDOCRINE DISEASE

Secretion rates. The best measure of adrenal cortical function is provided by measurement of secretion rates by the adrenal gland of cortisol and of aldosterone after injecting hormone labelled with radioactive isotope.

Congenital adrenal hyperplasia

In the adrenogenital syndrome due to congenital adrenal hyperplasia there is a congenital deficiency of 21-hydroxylase causing a block in the synthesis of cortisol with the appearance of abnormal quantities of 17-hydroxyprogesterone and its metabolite, pregnanetriol (Fig. 24). The deficiency of cortisol leads to over-production of ACTH, resulting in adrenal hyperplasia with over-production of the adrenal androgens and other intermediate compounds and their metabolites. The estimation of pregnanetriol in urine is important in the diagnosis and treatment of the syndrome. In some cases there is a deficiency of the 11-hydroxylase leading to an excessive production of the mineralocorticoid-compound S which can lead to hypertension. This compound and pregnanetriol are typical 17-hydroxycorticosteroids; a normal or even raised excretion of 17-hydroxycorticosteroids may therefore be observed, even though there is a deficiency of cortisol. In others there is a deficiency of 3β-hydroxy steroid dehydrogenase leading to the excessive production and excretion of pregnenolone and its metabolites. Separate estimation of these steroids may be thus important in this kind of patient. The metabolites of neither compound S nor of 17-hydroxyprogesterone are oxygenated in the 11-position, and it is often simpler and more convenient to estimate the 11-oxygenation index, i.e., the fraction of the metabolites which are oxygenated in the 11 position. This is normally about 0·2 and rises to above 2 in congenital adrenal hyperplasia. This does not require 24-hour collections of urine, and is particularly convenient in the investigation of infants. The following investigations are therefore of importance in the diagnosis of this condition.

Steroid investigations in congenital adrenal hyperplasia

(1) Urinary 17-OHCS—raised (mainly due to cortisol precursors).
(2) 17-Oxosteroids—raised.

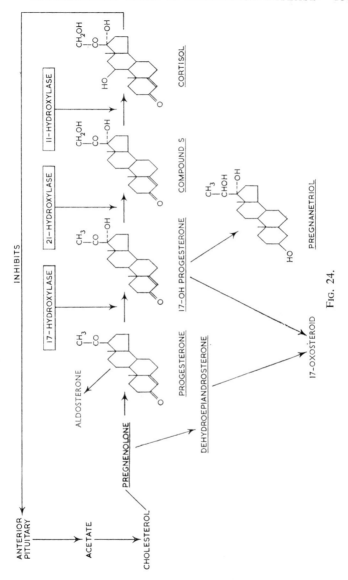

Fig. 24.

(3) Plasma cortisol—normal or low.
(4) Raised urinary 17-OHCS and 17-OS easily suppressed by dexamethasone.
(5) Urinary pregnanetriol—raised in 21-hydroxylase deficiency. Urinary tetrahydro S raised in 11-hydroxylase deficiency; both suppressed by dexamethasone.
(6) 11-Oxygenation index raised.

Many of these patients respond admirably to corticosteroid therapy, which suppresses the secretion of ACTH by the pituitary gland and reduces the excretion of virilizing steroids. The effectiveness of treatment should be controlled by measurements of 17-oxosteroids.

Cushing's syndrome

Cushing's syndrome is a condition associated with adrenal hyperactivity. This may be due to an adrenal tumour or to hyperplasia, secondary to tumour or hyperplasia of the basophil cells of the pituitary. The clinical features of Cushing's syndrome are simulated by many patients receiving treatment with ACTH, adrenal cortical hormones or their synthetic analogues. Occasionally non-endocrine tumours secrete ACTH or an ACTH-like substance, causing a very rapidly developing form of Cushing's syndrome; this is sometimes recognized in its early stages by a severe hypokalæmic alkalosis. The laboratory findings are ideally, but not always, as follows:

Autonomous adrenal tissue

(1) raised urinary 17-OHCS unaffected by ACTH or dexamethasone;
(2) raised plasma cortisol without diurnal rhythm;
(3) raised cortisol secretion rate.

Adrenal hyperplasia

(1) raised 17-OHCS with exaggerated response to ACTH but not suppressed with dexamethasone 2 mg/day but suppressed by 6–8 mg/day;
(2) raised plasma cortisol with altered or absent diurnal rhythm;
(3) raised cortisol secretion rate.

Adrenal carcinoma

Malignant tumours of the adrenal cortex are usually associated with abnormal biosynthesis of steroid hormones leading to excessive secretion of steroid hormones, with their precursors and metabolites. Initially these will be recognized by an excessive excretion of 17-oxosteroids and/or 17-hydroxycorticosteroids. The amounts are usually greater than are found in hyperplasia; their identity varies from patient to patient and requires special investigation.

Addison's disease

In Addison's disease the excretion of neutral 17-oxosteroids and of 17-hydroxycorticosteroids is low, but this is not the most satisfactory test for the disease. When there is a deficiency of adrenal cortical hormones in the body there is a deficiency in the reabsorption of sodium ions in the renal tubules, and the body becomes sodium deficient with all its attendant sequelæ. In addition, there is a shift of sodium from the extracellular fluid into the intracellular fluid and potassium in the reverse direction. A crisis in Addison's disease is thus a very severe form of salt deficiency (see Chapter IV) and is a medical emergency which requires immediate treatment, the diagnosis being subsequently confirmed by the low plasma sodium and chloride, the frequently raised plasma potassium and the raised blood urea. Treatment of such a crisis involves replenishment of the sodium by intravenous administration of 0·9% sodium chloride with added glucose, together with replacement therapy of the deficient adrenocortical hormones. The low sodium content of the plasma (and extracellular fluid) may in part be due to passage of sodium ions into the intracellular fluid. Reversal of this process by intravenous cortisol may lead to considerable clinical recovery before sodium therapy has had time to restore the chronic renal loss of sodium. The plasma cortisol should be measured before treatment, which must not be delayed until the result is available.

The diagnosis of chronic Addison's disease sometimes presents a serious problem. Analysis of the blood electrolytes is often helpful, but not always, because loss of sodium by the body may be paralleled by loss of water. The concentration of sodium in the plasma therefore may remain practically within normal limits, despite the presence of severe sodium deficiency. Patients in this condition not only excrete excessive amounts of sodium (mainly as chloride) but there is also deficiency in the excretion of urea and water. The minimum requirements for diagnosis of chronic Addison's disease are: (1) low or absent 17-hydroxycorticosteroids in urine and/or low or absent plasma cortisol; (2) deficient response of urinary 17-hydroxycorticosteroids or plasma cortisol to the synthetic polypeptide Synacthen, or ACTH.

Aldosteronism. The urinary excretion of aldosterone exceeds 20 μg/day in primary aldosteronism due to a tumour *as well* as in secondary aldosteronism associated with the nephrotic syndrome, cardiac failure or hepatic cirrhosis. For the diagnosis of an aldosterone-secreting tumour it may be necessary to estimate the aldosterone secretion rate by isotopic methods, but more recently the association of a high-aldosterone excretion with a low-plasma renin has been regarded as diagnostic of the condition.

The adrenal medulla

Phæochromocytoma of the adrenal medulla or of accessory chromaffin tissue is characterized by the intermittent secretion of adrenalin and noradrenalin and their metabolites. Biological methods are available for the determination of these two hormones in blood and urine, but are best carried out in laboratories specializing in this field. The chemical estimations of the metabolite "vanillylmandelic acid" (VMA)* or of metanephrin and normetanephrin in urine are useful as screening tests.

The pituitary

Immunochemical methods are available in specialized departments for the estimation of growth hormone, TSH and LATS. There are still difficulties for ACTH and the pituitary gonadotrophins (see below). The polypeptide hormones of the posterior pituitary have hitherto required biological methods, but recently these, too, may be estimated immunologically.

Panhypopituitarism. In panhypopituitarism there is a deficiency of gonadal, adrenal and thyroid secretions, secondary to a lesion of the anterior pituitary often due to necrosis after

* More correctly 4-hydroxy-3-methoxy mandelic acid (HMMA).

severe puerperal hæmorrhage. The excretion of 17-oxosteroids is very low. The PBI and radio-iodine tests reveal hypothyroidism, which, however, responds to injection of TSH. Although the adrenal and gonadal insufficiency is associated with a virtual absence of neutral 17-oxosteroids from the urine, the other functions of the adrenal cortex do not appear to be so severely reduced, for a deficiency of sodium and other electrolyte disturbances are uncommon, and crises as seen in Addison's disease do not occur. However, that adrenal insufficiency is present is demonstrable by the extreme insulin sensitivity of these cases. Plasma cortisol and 17-hydroxycorticosteroid excretion in panhypopituitarism are low and do not respond to metopirone, but are increased after Synacthen and ACTH; this response may sometimes require treatment with these hormones for two or three days.

Hyperpituitarism. The clinical and biochemical features of basophil adenoma are those of Cushing's disease (see above). Gigantism and acromegaly occur due to excessive growth hormone production by the anterior pituitary or by eosinophil adenoma. The diagnosis usually rests upon X-ray or clinical evidence. The destruction of the other elements of the pituitary leads to a superimposed pituitary myxœdema and hypogonadism with a diminished 17-oxosteroid excretion, During the acute phases of the disease the plasma inorganic phosphate is high, and the blood contains an excess of both growth hormone and of insulin. The growth hormone presumably stimulates the pancreas to secrete additional insulin, which, however, is insufficient to antagonize completely the diabetogenic activity of the growth hormone and there results a characteristic form of diabetes.

The gonads

The testes and ovaries both produce steroid hormones—the former mainly testosterone and the latter mainly œstradiol and œstrone; many of the pathways of biosynthesis and metabolism and the secreted hormones and excreted metabolites are the same as those of the adrenal hormones. All three endocrine glands synthesize androstenedione and DHA (see p. 146) from precursors identical with those of the adrenal cortical hormones. Œstrogens are synthesized with androgens as intermediate compounds (see Fig. 25). The conversion of androstenedione and DHA to œstrogen is the major pathway in the ovaries; in contrast, in the testes the conversion of the androgen to œstrogen is minimal, so that the major androgens secreted by the testes are testosterone and androstenedione. The quantitative

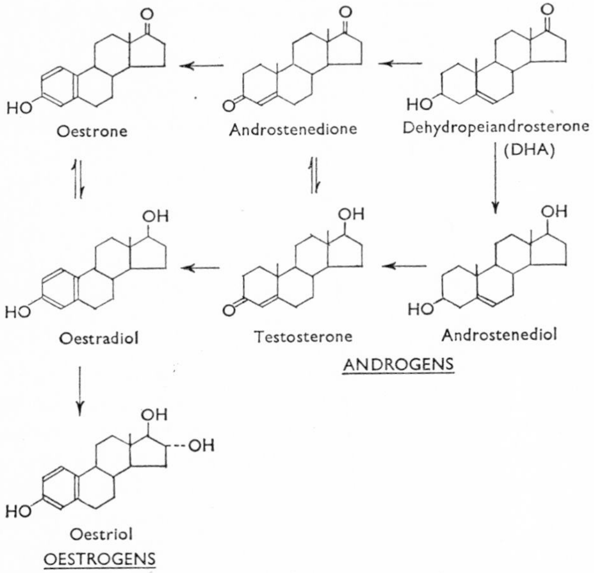

FIG. 25. Biosynthesis and metabolism of androgens and œstrogens.

aspects of these interconversions are still being investigated, and other pathways of biosynthesis may prove significant.

Hirsutism in women is associated with an abnormally high concentration of testosterone (which in women is normally very low) in the peripheral blood; in some this testosterone is derived mainly from the ovaries, in others from the adrenals. In the Stein–Leventhal syndrome, infertility, amenorrhœa and hirsutes are associated with fibrocystic ovaries, in which there is a deficient conversion of androgen to œstrogens.

Menstruation. The menstrual cycle is controlled by the excretion of follicle-stimulating (FSH) and luteinizing (LH) hormones of the pituitary, and the first half of the cycle is associated with a peak in the production of œstrogens from the maturing follicles. During the second half of the cycle, after ovulation, the development of the corpus luteum under the influence of the LH is associated with progesterone production and a second peak of œstrogens. In the absence of pregnancy there is a precipitous fall in the œstrogens, and the uterine decidua is shed.

Pregnancy. If pregnancy occurs, the excretion of œstrogens, progesterone and LH increases, and ultimately the fæto-placental unit takes over their production. After the third month the LH excretion falls somewhat.

More than ten other closely related œstrogens or their metabolites are now recognized, and the estimation of the œstrogens is highly complex and undertaken only in specialized laboratories; such investigations provide as yet little indication for treatment of disorders of menstruation; estimations of œstriol in late pregnancy are believed to provide an index of placental function and fœtal viability.

Progesterone, the progestational hormone of the corpus luteum and placenta, and a precursor of all the adrenal cortical hormones, is not excreted in the urine as such, but is metabolized to a variety of other products, including pregnanediol, which, however, represents only a small fraction of the secreted progesterone. Excellent methods are now available for the estimation of progesterone and its metabolites by gas chromatography, but are not yet suitable for routine use. There is evidence that some patients subject to habitual abortion are progesterone deficient, and measurement of pregnanediol excretion during early stages of pregnancy has been used to indicate the necessity for treatment with progesterone. The excretion of pregnanediol can now be estimated during the latter half of the menstrual cycle, but its importance in the diagnosis and treatment of disorders of menstruation has not yet been clarified.

Gonadotrophins

The gonadotrophins are protein hormones, and immunochemical methods have almost replaced the biological assays for the detection and determination of chorionic gonadotrophin (HCG). However, difficulties have been encountered with the immuno-assay of FSH and LH.

Follicle-stimulating hormone (FSH). FSH must at present be estimated in urine by biological assay and is increased in primary ovarian deficiency, including that of the menopause. No detectable FSH is found in pituitary deficiency, including that of complete hypophysectomy.

Chorionic gonadotrophin (HCG). HCG excreted during pregnancy and by hydatidiform mole and chorioncarcinoma is now detected qualitatively by a variety of immunological tests instead of by biological methods involving the Friedman, Aschheim–Zondek or Hogben tests. An accurate immunological assay of HCG is of importance in diagnosing and following the treatment of the last two conditions.

Hypogonadism. The excretion of the neutral 17-oxosteroids is less than normal in hypogonadism in males, in Addison's disease and in Simmonds' disease, as well as in severe debilitating illness and old age. The wide range of normal values and of proportions of these compounds derived from the adrenal cortex limits the value of their determination in hypogonadism.

Tumours of the gonads

Tumours and abnormalities of the testes are seldom accompanied by an increased excretion of the neutral 17-oxosteroids. A number of testicular tumours secrete HCG, but our knowledge of these tumours is insufficient to provide a satisfactory classification. Arrhenoblastomas and granulosa-cell tumours of the ovary secrete large quantities of oxosteroids and œstrogens respectively.

FURTHER READING

Association of Clinical Pathologists Symposium on Thyroid Gland. 1967. *J. Clin. Path. Suppl.* **20**, 309.

CONDLIFFE, P. G. and ROBBINS, J. 1967. "Pituitary Thyroid-Stimulating Hormone and other Thyroid-Stimulating Substances", in *Hormones in Blood*, Volume 1. 2nd edition. Edited by C. H. Gray and A. L. Bacharach. Academic Press; London, New York.

LORAINE, J. A. 1966. "The Clinical Application of Hormone Assay." 2nd edition. Livingstone; Edinburgh.

PRUNTY, F. T. G. 1967. "Hirsutism, Virilism and Apparent Virilism and Their Gonadal Relationship." Part I. *J. Endocrin*, **38**, 85–103. Part II. *J. Endocrin.* **38**, 203–227.

PRUNTY, F. T. G. 1964. "Chemistry and Treatment of Adrenocortical Diseases." Charles C. Thomas: Springfield, Ill.

ROBBINS, J. and BALL, J. E. 1967. "The Iodine-Containing Hormones", in *Hormones in Blood*, Volume 1. 2nd Edition. Edited by C. H. Gray and A. L. Bacharach. Academic Press; London, New York.

WRIGHT, A. D. and TAYLOR, K. W. 1967. "Immuno-assay of Hormones", in *Hormones in Blood*, Volume 1. 2nd Edition. Edited by C. H. Gray and A. L. Bacharach. Academic Press; London, New York.

CHAPTER XI

BIOCHEMICAL GENETICS

DURING the first quarter of the twentieth century Archibald Garrod drew attention to the inborn errors of metabolism, rare diseases each characterized by a specific block in a biochemical pathway and resulting in the disposal, by some alternative route of a metabolite immediately preceding that metabolic block. Garrod recognized that while some of these abnormalities, such as pentosuria, were without injurious effects, others, including congenital porphyria, were definitely harmful. He then postulated that these abnormalities were due to an inheritable enzyme deficiency. Since then many such anomalies of metabolism have been discovered, and the mechanisms of their inheritance have been investigated and fitted into a general theory of biochemical genetics. This theory also provides an understanding of the mechanism of inheritance of certain normal characteristics, such as the blood groups and the types of hæmoglobin.

The detailed study of mutations induced in micro-organisms led to the one gene–one enzyme hypothesis, according to which the chromosomal genes control the chemistry of the cell by directing the production of enzymes. Each gene is responsible for controlling the synthesis of a single enzyme; absence, abnormality or deficiency of that gene will lead to an absence, abnormality or deficiency of the corresponding enzyme. There results either a block in a metabolic pathway or of a synthesis of some specific molecule, such as a hæmoglobin or a blood group specific substance.

Except for the gametes and their immediate precursor cells, all normal cells are diploid and contain a paired set of chromosomes, one of maternal and the other of paternal origin. In man there are thousands of genes arranged uniquely along the length of each chromosome. Each gene is responsible for the synthesis of an enzyme, and if any specific pair of genes are of the same kind the individual is homozygous in respect of this gene; if the two members of the pair of genes are different the individual is said to be heterozygous.* With recessive characteristics this may have important quantitative implications on enzyme synthesis. Thus individuals homozygous in respect of the sickle cell gene will form sickle-cell hæmoglobin instead of the normal adult variety. Because of the specific enzymic deficiency, a single amino acid has been changed in the hæmoglobin molecule so that the reduced pigment readily crystallizes in the circulating red cell, which acquires the characteristic shape and is readily hæmolysed. In individuals heterozygous in respect of the sickle-cell gene, only half of the hæmoglobin is of the sickle cell variety, and these individuals suffer from "the sickle cell trait", which becomes manifest clinically only under certain precipitating conditions. The plasma of some individuals contains immeasurable amounts of cholinesterase, and they are abnormally susceptible to suxamethonium, a relaxing agent often used in modern anæsthesia. Suxamethonium is normally broken down by cholinesterase. These individuals must be homozygous in respect of the gene responsible for the synthesis of this cholinesterase. The plasma from some members of their families contains only half the normal amount of the cholinesterase; these individuals must therefore be heterozygous. Beyond reminding the reader of the penetrance of genes, which may be of various degrees, it is beyond the scope of this book to consider genetics and heredity in detail.

Blood group specificity. The differences between the main blood groups are due to the presence of inheritable antigenic mucopolysaccharides in the red-cell surface. These antigens are synthesized by enzymes, the formation of which is under the control of pairs of genes or alleles. These are also responsible for the synthesis of similar group-specific mucopolysaccharides which are major constituents of the mucous secretions of the body.

The blood proteins. More than 29 variants of human hæmoglobin have been characterized by zone electrophoresis on paper, or in starch or agar gel. Free boundary electrophoresis and chromatography have been of little value in their separation.

* Genes which although different can occupy the same site on the chromosome are called alleles.

They show differences in their solubility in different solvents as revealed by discontinuity in their salting-out curves. They also differ from one another in their solubility, often in their oxygen dissociation and the readiness with which they are denatured. They show immunological specificity, and their production appears to be due to the presence of alleles in the gene responsible for the synthesis of the hæmoglobin.

Normal adult hæmoglobin is mainly hæmoglobin A with a minor component hæmoglobin A_2. Occasionally this last component is increased. Fœtal hæmoglobin is probably heterogeneous and is more slowly denatured by alkali than adult hæmoglobin; this formerly provided the basis of a method for its determination. Sickle-cell hæmoglobin (see above) has been most completely studied. It separates well from hæmoglobin A even with paper electrophoresis, and even better with other techniques. Other hæmoglobins have been given the letters of the alphabet, the names of the family or the place in which they have first been found. Ingram has investigated the specific amino acid sequences in many of these hæmoglobins, and these sequences and their genetic inheritance have been of particular importance to the study of biochemical genetics. Frequently these variations are due to an alteration in a single amino acid or group of amino acids in the molecule, the synthesis of which is under the control of a single gene.

Some genetically determined abnormalities of the plasma proteins are considered in Chapter VI.

Genetically determined variants of enzymes. We have seen that suxamethonium sensitivity is due to a deficiency of cholinesterase activity and glucose-6-phosphate dehydrogenase deficiency which occurs with an appreciable incidence in particular population groups may lead to sensitivity of the red cells to broad beans and to primaquine. Such enzyme diversity has proved to be much more common than had been supposed. Detailed investigation of the plasma from subjects with suxamethonium sensitivity has revealed no less than four different cholinesterases differing from one another and from those found in normal individuals in their sensitivity to inhibiting agents. The precise pattern of cholinesterase activity is determined by four allelic genes which can pair in ten different ways to give homozygous or heterozygous individuals.

The electrophoretic patterns of other enzymes in the body have been shown to be highly specific in individuals. Thus, acid phosphatase and phosphoglucomutase of red cells, serum alkaline phosphatase, placental alkaline phosphatase and the adenylate kinase of red cells and of muscle prove, on electrophoresis under suitable conditions, to be mixtures of iso-enzymes, the precise proportions of which appear to be genetically determined. The adenylate kinase can be separated into iso-enzymes which show that its synthesis is determined by two pairs of alleles. There proves to be a marked genetic individuality in the physico-chemical properties of the body enzymes. These genetically determined variants of enzymes together with similarly determined variants of blood groups, hæmoglobin, haptoglobin, etc., must provide a biochemical basis for individuality and must necessitate a careful reinvestigation of the application of iso-enzyme patterns to clinical diagnosis (see p. 171).

Disturbances of amino acid metabolism

Alkaptanuria and albinism. In alkaptanuria and albinism, two of Garrod's original inborn errors of metabolism, there are blocks in the metabolism of phenylalanine and tyrosine respectively. In the former condition the metabolism of these amino acids is blocked at the homogentisic acid stage; this acid is therefore excreted. The urine darkens on standing and is strongly reducing. Except for a tendency to rheumatic disease, presumably related to the deposition of pigment in the joint cartilages, the condition is harmless. In albinism there is deficiency of the enzymes responsible for melanin formation from tyrosine.

Phenylketonuria. In about half per cent of mental defectives the mental disturbance is related to the excretion of phenylpyruvic acid, together with phenylalanine, phenyllactic acid, phenylacetic acid and its glutamine conjugate. Of these, phenylpyruvic acid is the more readily detected (see page 213). These patients have an inherited deficiency of the enzyme responsible for forming tyrosine from phenylalanine which is present in the plasma in greatly increased concentrations. These patients are now being treated with phenylalanine-free diets as early in their lives as possible, but the success of this therapy will depend upon whether the mental deficiency is caused by the toxic action of phenylalanine itself or its derivatives, or whether the mental disturbance and the biochemical defect are both secondary to some more fundamental abnormality.

The site of the metabolic defect is shown in Fig. 26, which shows the alternative metabolic pathways which become important in the condition. Fig. 26 also shows the probable metabolic blocks responsible for alkaptonuria, albinism and certain forms of cretinism.

Amino acidurias. Amino acids which are normally present in the plasma and extracellular fluid compartment are generally almost completely absorbed from the glomerular filtrate by tubular absorption. In liver disease as well as in some inheritable biochemical disorders there is a production of amino acids in excess which exceeds the normal reabsorptive capacity of the renal tubules, leading to an overflow "amino aciduria". In other congenital abnormalities the mechanism is renal and there is an abnormality of the tubular absorption of certain amino acids or groups of amino acids which share a common tubular reabsorptive mechanism. In overflow amino aciduria the plasma concentrations are usually raised, but since the increase may affect only one or few of the amino acids present, this is not always easy to demonstrate. In the renal amino acidurias plasma concentration may be normal or low.

The Fanconi syndrome, a condition which is one of the causes of failure to thrive in infancy, is characterized by a marked amino aciduria involving many amino acids, renal glycosuria and a low plasma inorganic phosphate, associated in childhood with a vitamin D-resistant rickets or with osteomalacia in adult life. In addition, there is often potassium loss and hyperchlorӕmic acidosis due to failure of the renal tubules to excrete hydrogen ions. In some patients with this condition the abnormal renal tubular reabsorption may be limited to only one or more of the amino acids. The condition is best explained as renal in origin, i.e., a general defect of tubular reabsorption, but this does not provide an explanation of the deposition of cystine crystals, which can be deposited throughout the tissues in one of the commonest forms of the disease. This cystinosis of the Fanconi syndrome and its variants differs markedly from cystinuria, in which there is no deposition of cystine in the

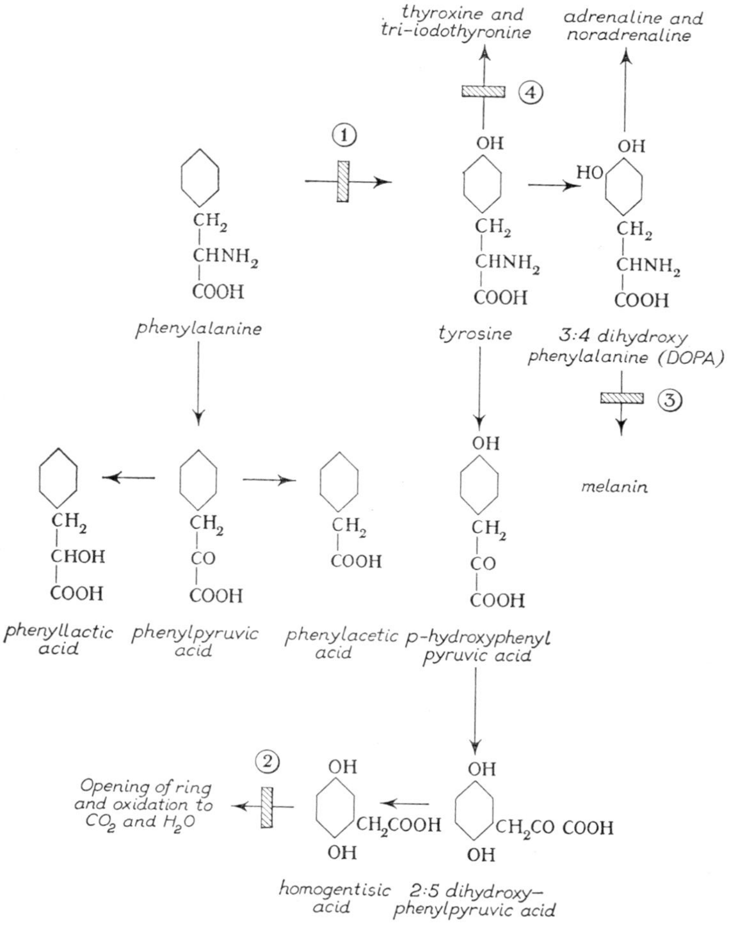

FIG. 26. Sites of Metabolic Blocks Responsible for: 1 phenylketonuria, 2 alkaptanuria, 3 albinism, 4 certain forms of cretinism.

tissues but a genetically determined deficiency in the tubular reabsorption of the amino acids cystine, lysine, arginine and ornithine. Of these four, the less-soluble cystine crystallizes, forming recurrent calculi. Other abnormalities of amino acid metabolism include the excretion of cystathionine and of arginino-succinic acid, an intermediate in the cycle responsible for the synthesis of urea. In Hartnup disease there is a pellagra-like rash with cerebellar damage associated with amino aciduria and indolylacetic aciduria. "Maple syrup urine disease" is a fatal neurological disease affecting young infants, and is associated with the excretion of valine, leucine and isoleucine, while a number of individuals have been found who apparently without harmful effects, excrete increased amounts of β-amino-isobutyric acid.

Disturbances of carbohydrate metabolism

Renal glycosuria (see pp. 17 and 133) is usually a congenital abnormality of glucose reabsorption, but may sometimes occur in primary renal disease. The genetically determined form may be due to either diminished glucose reabsorption in *all* the nephrons or to a wider range than normal of glucose reabsorption in the various nephrons. In this last form some nephrons will thus have abnormally high absorption, while others have an abnormally low absorption, resulting in a mean threshold which is normal.

Fructosuria and pentosuria. These are normally harmless abnormalities related to a lowered renal tubular reabsorption of the fructose and of the pentoses of the diet. However, one type of pentosuria, 1-xylosuria, is due to an enzymic deficiency interfering with glucuronic acid metabolism.

Congenital galactosæmia is important and due to a deficiency of the gene controlling the synthesis of the enzyme galactose-1-phosphate uridyl transferase concerned with the conversion of galactose to glucose. The enzyme brings about the reaction of galactose-1-phosphate with uridine diphosphoglucose to form glucose-1-phosphate and uridine diphosphogalactose, which is readily converted to the corresponding glucose compound (see Fig. 27). This enzyme is virtually absent in the erythrocytes and liver of galactosæmic patients, and its determination is essential for diagnosis of the condition. Patients with the disease are children who fail to thrive and the intra- and extra-cellular accumulation of galactose-1-phosphate causes mental retardation, severe liver damage and cataracts. There is also a secondary amino aciduria. It is likely that normal growth and development can occur if a galactose-free régime is started early enough. Early diagnosis is therefore essential (see page 208).

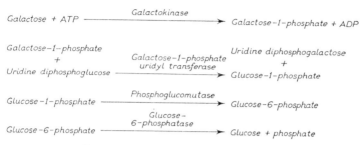

FIG. 27. Pathway of galactose to glucose.

Glycogen storage disease (Von Giercke's disease) occurs in a number of different clinical forms according to which of the enzymes responsible for degrading glycogen is deficient (Fig. 28). In the usual form of the disease there is hepatomegaly, failure to thrive, with attacks of hypoglycæmia and ketosis; it affects the liver and kidneys and is due to a deficiency of glucose-6-phosphatase. Another form affecting also the skeletal and heart muscles is associated with a deficiency of amylo-1-6-glucosidase, the enzyme responsible for debranching the glycogen molecule. In a rare form of the disease an abnormal glycogen is deposited and is due to a deficiency of the enzyme responsible for introducing the branches into the molecules of glycogen.

Red-cell abnormalities associated with glucose-6-phosphate dehydrogenase deficiency (primaquine sensitivity, favism). Deficiency of glucose-6-phosphate dehydrogenase, more marked in the homozygous than in the heterozygous individual, causes an abnormal sensitivity of the red cells to drugs such as primaquine and other drugs, as well as to the broad bean *fava vicina*. Homozygous subjects are especially sensitive, even to the pollen. The mechanism of the hæmolytic process is not yet fully understood, but the integrity of the red cell depends upon the maintenance of the glutathione in the reduced form; if this mechanism fails hæmolysis takes place.

Methæmoglobinæmia. The hæmoglobin of the circulating red cells is continuously being converted to its oxidized form methæmoglobin, but this is normally rapidly reduced and maintained in very low concentrations by a series of enzymes. In certain subjects one of the components of the enzyme system, diaphorase, is deficient, and an excess circulating hæmoglobin is maintained in the oxidized condition. This results in cyanosis, which, however, can be relieved therapeutically by ascorbic acid or by methylene blue. Methæmoglobinæmia can also occur after the administration of drugs such as acetanilide, antipyrine, phenacetin or nitrite.

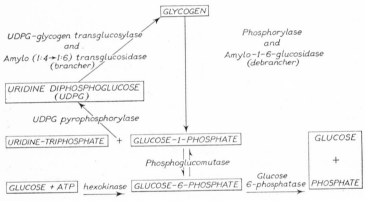

FIG. 28. Glycogen synthesis and degradation.

Other inheritable metabolic diseases

The porphyrias. In congenital porphyria, one of Garrod's original inborn errors, there is an enzymic abnormality of porphyrin biosynthesis which, instead of being directed almost specifically to the uroporphyrin III required for the synthesis of protoporphyrin (the prosthetic group of hæmoglobin), produces an equal quantity of a uroporphyrin I unsuitable for hæmoglobin synthesis. This uroporphyrin I causes severe photosensitivity and is excreted in the urine.

A second form of porphyria, acute intermittent porphyria, is associated with an enzymic deficiency at an earlier stage in porphyrin biosynthesis (p. 192). There is a greatly increased excretion of δ-aminolævulic acid and porphobilinogen, the colourless precursors of the porphyrins. Porphobilinogen is readily detected in urine by Ehrlich's aldehyde reagent, which gives with it a red colour, even though Schlesinger's test is negative. The urine is often colourless, but the porphobilinogen is rapidly converted to a mixture of red and red-brown porphyrins and porphyrin-like pigments on standing. In this disease there are abdominal colics and nervous-system lesions but no photosensitivity. Other types of porphyria exist which resemble mild forms of these two types in the same individual, but in these the enzymic deficiency is not yet known.

Congenital hyperbilirubinæmia. In some forms of this condition there is a deficiency of uridine diphosphoglucuronyl transferase, the enzyme responsible for transferring the glucuronyl group to bilirubin. This enzymic deficiency is inheritable and provides one of the examples of non-hæmolytic pre-hepatic jaundice (see pp. 67 and 74).

Other hereditary enzyme defects. It is probable that primary hæmochromatosis, hyperuricæmia, essential hypercholesterolæmia, primary oxalaluria and hypophosphatasia are other examples of genetically determined enzyme deficiencies. Congenital adrenal hyperplasia has recently been shown to be due to a deficiency of the enzyme responsible for introducing certain hydroxyl groups into the progesterone molecule in the synthesis of cortisol. Other hydroxylations of progesterone can take place, but as described in Chapter X, the failure to synthesize cortisol leads to excessive production of ACTH. This stimulates overproduction of the 17-oxosteroids and of the progesterone metabolites, pregnanetriol, and pregnanetriolone which are excreted in the urine as glucuronides.

Further Reading

CARTER, C. O. 1967. "An Introduction to Biochemical Genetics." *J. Roy. Coll. Phys. of London*, **1**, 167.

HARGREAVES, T. 1963. "Inherited Enzyme Defects: A Review." *J. clin. Path.*, **16**, 293.

HARRIS, H. 1964. "The Genetics of Serum Cholinesterase 'Deficiency' in Relation to Suxamethonium Apnœa". *Proc. Roy. Soc. Med.*, **57**, 503.

"Human Genetics", ed. A. C. Stevenson, *Brit. med. Bull.*, 1961, **17**, No. 3.

CHAPTER XII

CLINICAL ENZYMOLOGY

THE majority of chemical changes in the body are mediated by enzymes, which are essential constituents of the cells and may either remain intracellular or pass out of the cells into the extracellular fluid, into the bloodstream or into the secretions. When certain enzymes are deficient, as in the hereditary anomalies discussed in Chapter XI, metabolic abnormalities develop and may be associated with overt clinical disease. In some instances the enzyme deficiencies were postulated to explain the abnormality even before the precise enzyme-controlled metabolic pathway was known.

Disease caused by trauma to the tissues by bacterial or viral invasion or brought about by mechanical or chemical means might be associated with derangement of enzyme activity. However, the enzyme changes resulting from such trauma are only just beginning to be investigated. Thus the histological changes after carbon tetrachloride poisoning have been shown to be preceded by a failure in the co-ordination of activities of the enzymes of the tricarboxylic acid cycle.

The role of deficiencies of the digestive enzymes in certain diseases have been known for nearly half a century, and the importance of measurements of plasma alkaline and acid phosphatase in the diagnosis of liver and bone diseases recognized for twenty-five years. However, great advances in clinical enzymology have been made in the last ten years, mainly due not only to more detailed knowledge of enzyme mechanisms but also to the more ready availability to clinical laboratories of purified coenzymes and specialized equipment such as the spectrophotometer. Such laboratories can now undertake enzyme measurements previously confined to laboratories concerned more with fundamental biochemistry than with clinical diagnosis.

A number of enzymes are present in normal blood. They may have been released from some intracellular site because the cell membranes of certain cells are permeable to them or they may arise from the general "wear and tear" of cells. The activity of these plasma enzymes is normally kept low by inactivation or by excretion. Thus, phosphatase is excreted in the bile while amylase is excreted in the urine. In some diseases the enzyme activity in plasma may be increased by the diversion of the route of excretion, as in obstructive jaundice when alkaline phosphatase is no longer excreted in the bile. The acid phosphatase of plasma is increased when carcinoma of the prostate metastasizes in bone because the tumour cells secrete the enzyme directly into the bloodstream instead of into the prostatic tubules.

Increased enzyme activity of plasma can also be due to increased synthesis of the enzyme, and this, together with passage from the intracellular site to the bloodstream, probably accounts for the increased alkaline phosphatase in the plasma associated with certain bone diseases characterized by increased osteoblastic activity.

Changed permeability of cell membranes by inflammation can permit release of intracellular enzymes, but it is when there is necrosis of a tissue that large amounts of enzyme pass into the bloodstream. If the tissue affected has a specific pattern of enzymes this pattern is reflected in the changes which occur in the blood plasma. Thus heart muscle, liver cells and skeletal muscles are particularly rich in glutamic oxaloacetic transaminase, and necrosis of one or other of these tissues is followed by an increased activity of this enzyme in the blood.

Iso-enzymes. These are proteins of similar enzymic activity which differ in their physico-chemical properties and may be separated by some physical method such as electrophoresis or chromatography. Lactic dehydrogenase and many other enzyme activities are often characteristic of the particular organ from which they are derived, and the pattern of iso-enzymes in plasma has been claimed to assist in indicating the particular tissue from which any excess enzyme has been derived. The recent work, however, showing genetically determined patterns of iso-enzymes in different groups of individuals (see Chapter XI), suggests that more detailed investigation of these iso-enzyme patterns for individual patients may need to be a good deal more thorough than has been the case.

Technical problems of enzyme assay in body fluids. The methods of clinical enzymology differ from those of other investigative methods of chemical pathology because all enzymes are proteins, and care must be taken during preparation of the plasma for analysis to avoid denaturation which can occur even as a result of mechanical handling. Certain anticoagulants must be avoided too, as they may cause inhibition of the enzyme. The plasma should be separated rapidly to avoid contamination with enzymes present in the red cells, and if analysis cannot be performed immediately the plasma should be stored at $-20°$ C.

The analytical methods, which provide a measure of enzyme activity and not of true concentration because of the complex nature of enzymes, need to be chosen with care. Enzyme activity is dependent on temperature, pH of the reaction mixture, the presence of optimal amounts of co-enzymes, supply of reactants, as well as upon the removal of reaction products. The activity may be enhanced or inhibited by drugs or by ions of metals of ubiquitous occurrence in the body. It is small wonder that methods have been developed, the sole merit of which is speed and convenience rather than accuracy of estimation. Many results of clinical enzymology are thus entirely empirical, and it is still essential for each laboratory to establish its own range of normal values.

International agreement has recently been reached that enzyme activities should be expressed as international units, i.e., as the number of micro-moles of substrate changed per minute; there is, however, as yet no internationally agreed temperature at which enzyme activity should be measured. Many enzyme activities have for many years been expressed in arbitrary units, often based upon the liberation of 1 mg of product in a certain period. The widely accepted King–Armstrong unit of phosphatase activity is defined as the number of mg of phenol liberated in 15 min. from phenyl phosphate under the conditions of the test. It will be many years before units such as these will be abandoned, but all new applications of clinical enzymology should make use of the international units.

The digestive enzymes. The digestive enzymes which are of routine significance are the amylase in serum and urine, pepsin in gastric juice and urine, amylase, lipase and trypsin in pancreatic juice and trypsin in the fæces of children. These have been considered in Chapter VIII. Other enzymes are secreted into the gastrointestinal tract, but as yet their measurements are of no routine importance.

The phosphatases. The phosphatases of clinical significance may be broadly divided into: (a) alkaline phosphatase; (b) acid phosphatase; (c) 5'-nucleotidase; and (d) glucose-6-phosphatase. The last two are highly specific in respect of the substrate upon which they act.

The importance of the measurement of the alkaline phosphatases in plasma for the differential diagnosis of hepatobiliary disease and of osteoblastic bone disease is described in Chapters V and VII respectively. The isoenzyme patterns of the alkaline phosphatases of bone and liver are different, and may occasionally prove of value in diagnosis.

5'-nucleotidase, specific for the hydrolysis of nucleotide pentose-5'-phosphate groups, is raised in obstructive jaundice and much less so in hepatogenous jaundice. It is not increased in many of the other conditions accompanied by an increase in the ordinary plasma alkaline phosphatase. It is claimed to be of importance in the diagnosis of neoplasms of liver and those of bone, since in malignant disease involving the liver the serum 5'-nucleotidase is considerably increased.

Glucose-6-phosphatase occurs almost entirely in the liver, and the analysis of biopsy specimens for this enzyme is essential for the diagnosis of one type of glycogen storage disease (see page 167).

The transaminases. The transaminases* are enzymes which transfer amino groups; the most readily estimated are glutamic oxaloacetic transaminase (GOT) and glutamic pyruvic transaminase (GPT), which transfer amino groups from glutamic acid to oxaloacetic acid and pyruvic acid respectively (Fig. 29). These enzymes are widely distributed in human tissues, the amount of GOT in the heart being particularly high, while there is less in the liver; the greatest amount of GPT is in the liver. The concentrations in the serum are normally low (GOT 5–30 Reitman–Frankel units, GPT 0–25 Reitman–Frankel units), but become greatly increased when there is necrosis of tissues containing these enzymes. Thus, after myocardial infarction in man the GOT increases significantly within 6–12 hours, attains a peak

* In future these enzymes will be known as amino-transferases.

as high as 500–600 units within 24–48 hours and returns to normal within 4–7 days. Serial estimations should therefore be carried out, preferably at 12, 24 and 48 hours after the acute attack, otherwise false negative results may be obtained. With minor infarctions the peak level may only reach the upper level of the normal range, in which case subsequent estimations may

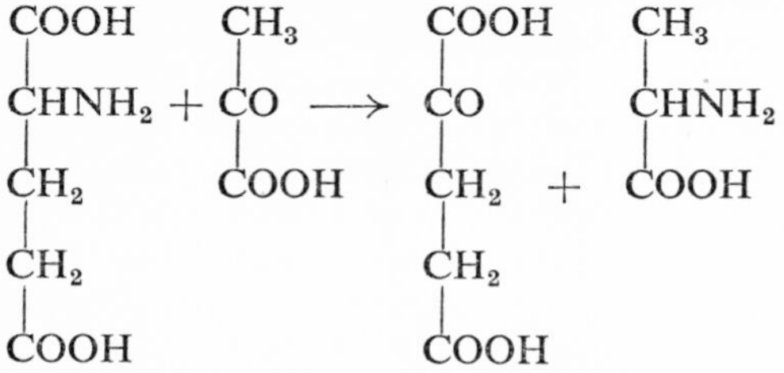

$$\begin{array}{cccc} \text{COOH} & \text{CH}_3 & \text{COOH} & \text{CH}_3 \\ | & | & | & | \\ \text{CHNH}_2 + \text{CO} \longrightarrow & \text{CO} & \text{CHNH}_2 \\ | & | & | & | \\ \text{CH}_2 & \text{COOH} & \text{CH}_2 + \text{COOH} \\ | & & | & \\ \text{CH}_2 & & \text{CH}_2 & \\ | & & | & \\ \text{COOH} & & \text{COOH} & \\ \text{Glutamic} & \text{Pyruvic} & \alpha\text{-Oxo} & \text{Alanine} \\ \text{acid} & \text{acid} & \text{glutaric} & \\ & & \text{acid} & \end{array}$$

Glutamic–pyruvic transaminase

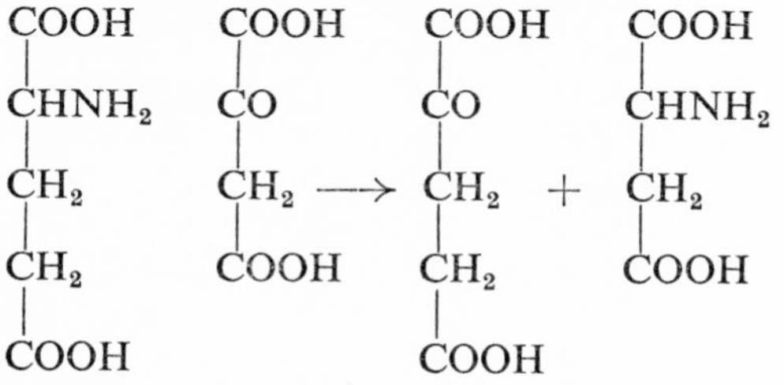

$$\begin{array}{cccc} \text{COOH} & \text{COOH} & \text{COOH} & \text{COOH} \\ | & | & | & | \\ \text{CHNH}_2 & \text{CO} & \text{CO} & \text{CHNH}_2 \\ | & | & | & | \\ \text{CH}_2 & \text{CH}_2 \longrightarrow & \text{CH}_2 + \text{CH}_2 \\ | & | & | & | \\ \text{CH}_2 & \text{COOH} & \text{CH}_2 & \text{COOH} \\ | & & | & \\ \text{COOH} & & \text{COOH} & \\ \text{Glutamic} & \text{Oxaloacetic} & \alpha\text{-Oxo} & \text{Aspartic} \\ \text{acid} & \text{acid} & \text{glutaric} & \text{acid} \\ & & \text{acid} & \end{array}$$

Glutamic-oxaloacetic transaminase

FIG. 29. The action of transaminases.

reveal a fall. The height of the peak of GOT in the serum may assist in the prognosis of the individual case. These estimations are specially important in cases of complicated coronary infarction when the clinical diagnosis is not secure. Cardiac failure after myocardial infarction can cause increases in both GOT and GPT because of the effect on liver function.

Estimations of serum transaminases are of value in the differential diagnosis of liver disease. Very high values (i.e., several thousands of units) are found in toxic liver disease, the concentrations returning to normal when exposure to the toxin is ended. With hepatitis, both enzymes are increased, and the increase in GOT is usually greater than that of GPT. These increases are maximal during the pre-icteric stage, when the patient is often most ill. GOT returns towards normal sooner than GPT, so that GOT becomes lower than GPT a few days after the appearance of jaundice. With recovery the concentrations return towards normal, but remain consistently high if the

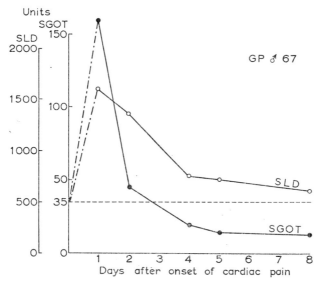

FIG. 30. Comparison of duration of SGOT and serum lactate dehydrogenase (SLD) in myocardial infarction. (By permission of Wilkinson.)

condition is progressing to sub-acute and chronic hepatitis. Secondary peaks may be associated with relapse. The GOT is also increased in cirrhosis and in neoplastic disease of the liver—the extent of the increase being roughly proportional to the amount of liver cells damaged in these conditions. Serum transaminase estimations are particularly of value in the differential diagnosis of jaundice in the newborn, in which it may be extremely important to distinguish as soon as possible between hæmolytic disease, physiological post-natal jaundice and obstruction of the biliary passages.

The dehydrogenases. This group of enzymes catalyse oxidation-reductions in the presence of a co-enzyme acting as hydrogen acceptor or donor. Those which have been alleged to be of value in clinical enzymology include lactate, isocitrate, malate, glutamate, glucose-6-phosphate and glutathione dehydrogenases and glutathione reductase.

Deficiency of glucose-6-phosphate dehydrogenase (which occurs mainly in the red cells) is associated with primaquine sensitivity and related hæmolytic disorders (see pages 167 and 196).

The serum lactic dehydrogenase rises for a short time after myocardial infarction and persists for about 7–12 days, remaining elevated after the SGOT has returned to normal (Fig. 30). The measurement is therefore of the greatest value when the patient has not been investigated until several days after the supposed myocardial infarction. Attempts have been made to correlate prognosis with the height of the maximum serum lactic dehydrogenase achieved, and maxima greater than 3,000 units are said to indicate invariably poor prognosis. In jaundice the elevation of serum lactic dehydrogenase activity is not so great as that of SGOT and SGPT. The serum lactic dehydrogenase is also elevated in many patients with disseminated forms of neoplastic disease, especially leukæmia, carcinomatosis, melanomatosis and Hodgkins' disease. The increase, however, is not invariable. Skeletal muscle contains large amounts of lactic dehydrogenase, and in progressive muscular dystrophy the serum level of the enzyme is considerably increased. Elevated levels of the enzyme are also found in untreated pernicious anæmia and renal disease.

Recently serum lactic dehydrogenase has been separated into isoenzymes by electrophoresis. The lactic dehydrogenases of the various tissues have characteristic patterns of isoenzymes, and it may well be that such investigations will prove to be of diagnostic value. Closely associated with the lactic dehydrogenase is an enzyme which reduces α-oxobutyrate, and its estimation is said to be of greater value than that of the serum lactic dehydrogenase in the investigation of cardiac infarction. The other dehydrogenases have been investigated in various conditions, but are not yet of clearly diagnostic value.

The glycolytic enzymes. Many of the glycolytic enzymes, including phosphoglucomutase, phosphohexoisomerase and

CLINICAL ENZYMOLOGY

aldolase, have been measured in diagnostic chemical pathology. In general, increases in their activity are to be expected in malignant disease, but as yet these measurements cannot be considered of major diagnostic importance. The serum aldolase, however, is greatly increased in progressive muscular dystrophy, presumably because skeletal muscle contains greater aldolase activity than any other tissue. The highest values are observed in the early stages of pseudohypertrophic type of progressive muscular dystrophy, but as the disease progresses the serum concentrations fall off, and in terminal stages reach normal values. The raised serum aldolase in this condition cannot be due solely to the breakdown of muscle, because similar increases do not occur in muscular atrophy and other neurogenic diseases. It has therefore been suggested that there is an abnormal permeability of the muscle fibres and that the actual muscle cells and myofibrils might have a shortened life span compared with normal muscle fibres.

Miscellaneous enzymes. Cholinesterase deficiency has been considered in relation to liver disease and to the genetically determined suxamethonium sensitivity in Chapters V and XI respectively. Measurements of pseudocholin esterase are also of importance in the diagnosis and treatment of poisoning by organophosphorus insecticides, which inhibit serum cholinesterase. Recently, creatine phosphokinase, the enzyme responsible for the phosphorylation of creatine, has been found to be increased in the serum of subjects with the Duchenne type of progressive muscular dystrophy and in patients with cardiac infarction.

The limitations of clinical enzymology. It has been widely accepted that when there is necrosis of a tissue the enzymes which may be present in large amounts in the cells are liberated and pass into the bloodstream. This mechanism explains the increase in certain enzymes in the bloodstream after coronary infarction, in hepatitis, in diseases of muscle and various other pathological conditions, perhaps even the increase in serum amylase in acute pancreatitis. The considerable overlap between the results found in the normal state and those found in pathological conditions and the overlap between the results obtained in various abnormal conditions has led to some disappointment in the general usefulness of clinical enzymology. The pattern of enzymes found in the plasma in disease is not necessarily the same as that of the tissue which has undergone necrosis. This may be due to the specific distribution of the enzymes in the various sub-cellular particles; leakage from necrosed or damaged cells may, therefore, depend upon the extent and rate of break-up of these sub-cellular particles. Moreover, many of the conditions associated with an abnormality may be accompanied by impairment of renal function, and this alone can lead to increased enzyme activity in the body fluids. The enzyme level in the plasma must depend upon the balance between its formation and degradation in the cell, its passage through the cell wall and perhaps other cell membranes, the rate of transfer to the extracellular fluid and circulating blood, as well as upon the rate of destruction in the blood and of removal by various organs and by elimination in the urine and bile. The rate of synthesis of enzymes is undoubtedly high in cells during regeneration, and this may be an important factor in liver necrosis, which is very soon accompanied by considerable regeneration of liver cells.

Disease is almost always a dynamic and changing condition. It is, of course, well recognized that the peak concentration of SGOT as a rule occurs at a well-defined time after coronary infarction, but this dynamic aspect of disease may account for the not always consistent changes which have been reported in plasma enzyme activities in various diseases. Great care should be taken that all enzyme estimations are carried out at the time optimal for expected abnormality to be observed.

Further Reading

BARON, D. N. 1966. "The Clinical Significance of Serum Enzyme Determination." *Abstracts of World Medicine*, **40**, 377.

DAWKINS, M. J. R. and REES, K. R. 1959. "A Biochemical Approach to Pathology." Edward Arnold, London.

LATNER, A. L. 1967. "Isoenzymes", in *Advances in Clinical Chemistry*, Vol. 9. Academic Press Inc.; New York.

WILKINSON, J. H. 1962. "An Introduction to Diagnostic Enzymology." Edward Arnold, London.

WILKINSON, J. H. 1965. "Isoenzymes." Spon's Biochemical Monographs.

CHAPTER XIII

SOME ASPECTS OF THE CHEMICAL PATHOLOGY OF THE NERVOUS SYSTEM

CERTAIN aspects of the chemical pathology of the central nervous system have been considered in other sections of this book. For example, the hyperexcitability of the central nervous system in tetany has been considered on page 97 and that due to cerebral œdema caused by water intoxication on page 56, while the chemical pathology of diabetic coma and of uræmia is described in detail in Chapters IX and I respectively. The genetic disturbances phenylketonuria, congenital galactosæmia, cretinism and acute porphyria associated with mental defect or nervous-system disease have been considered under the heading of Biochemical Genetics.

The chemical pathology of the cerebrospinal fluid. The mechanism of formation of the cerebrospinal fluid is complex, but as a first approximation it is usefully considered to be formed by mechanisms closely related to those concerned with the formation of the pleural, pericardial and peritoneal fluids. Ultrafiltration and osmotic reabsorption play a part, but there is no doubt that active secretion is also concerned. The condition of hydrocephalus in which excessive amounts of cerebrospinal fluid are formed must occur by mechanisms similar to those concerned with the formation of œdema and of ascites, etc., in other parts of the body.

The routine investigation of the cerebrospinal fluid. Chemical analysis should always accompany cytological and bacteriological examination of the C.S.F. because, though changes in concentration of many constituents parallel those of the blood plasma, this is not the case with glucose and proteins or substances such as bile pigment. Cytological and bacteriological findings will not be considered here, but a few comments concerning the chemical findings may be of value.

Presence of pigment. Cerebrospinal fluid is normally colourless, but sometimes may be pink or red from hæmorrhage due to accidental injury during lumbar puncture. Usually, but not always, the specimens collected last are less red and the quantity of blood is insufficient for the supernatant fluid to attain a yellow colour from admixture with plasma. If after centrifuging the red cells from C.S.F. the plasma is yellow and gives a van den Bergh test for bilirubin the hæmorrhage is most likely to be of pathological significance, unless lumbar puncture has been previously performed or jaundice has been present for some time. The fluid is sometimes yellow in pathological conditions such as meningitis.

Protein. The protein content of C.S.F. is normally less than 0·04 g/100 ml. It is increased in most pathological conditions because of alteration in the permeability of the blood-brain barrier. The greatest increases—i.e., up to 0·10–0·50 g/100 ml.— are found in various forms of meningitis and in the later stages of poliomyelitis and when a space-occupying lesion is present; lesser increases are found in cerebrospinal syphilis and in the early stages of poliomyelitis. When protein is present in considerable excess the concentration of fibrinogen is often sufficient to bring about a spontaneous coagulation on standing, a phenomenon in itself sufficient to diagnose tuberculous meningitis. When the protein content is raised qualitative tests for the presence of globulins often become positive. Certain neurological diseases are said to cause characteristic changes in the proportions of the various albumins and globulins which lead to abnormalities of the much-used Lange colloidal gold curve. The ability of serial dilutions of the C.S.F. to bring about a characteristic pattern of colour-changes in a colloidal gold sol is believed to be of value in diagnosis. A detailed account of these patterns is available in many standard text-books, and will not be presented here because of their empirical nature and because it is likely that electrophoresis on filter-paper or gel will permit a more satisfactory and less arbitrary indication of abnormality of partition of albumins and globulins.

Glucose. The concentration of glucose in the C.S.F. is dependent on: (1) the blood glucose concentration; (2) the utilization of glucose by any leucocytes that may be present; and (3) the utilization of any glucose by any organisms present. With normal blood sugar, C.S.F. glucose is reduced in meningococcal meningitis and even more so in tuberculous meningitis. Until recently a qualitative test was performed on the fluid; the glucose concentration was regarded as reduced when a negative result was obtained; quantitative determination is now necessary, because the recent use of streptomycin in the treatment of tuberculous meningitis has made determination important. Treatment is directed towards raising the C.S.F. glucose from values around 20–30 mg/100 ml. commonly found in this condition to the normal range of 70–80 mg/100 ml. Recent work has suggested that low concentrations of glucose in C.S.F. may be the cause of convulsions in infants without meningitis.

Chlorides. The chloride concentration of the C.S.F. is slightly higher than that of plasma. It was said to be specifically lowered in tuberculous meningitis, but it is now known that this was a reflection of the low plasma chloride that was common in this condition before streptomycin therapy. Confusion also occurs through expressing the concentration in mg sodium chloride/100 ml. with a normal range of 700–760 mg/100 ml. Since the sodium and chloride ions can vary independently of each other, the concentration should be expressed in mg chloride ion/100 ml. or better in mEq/l., in which case the normal ranges are 425–460 mg Cl/100 ml. or 120–130 mEq/l.

Vitamin deficiencies and the central nervous system. Deficiency of thiamine, whether due to frank deficiency in the diet or to the induced deficiency observed in chronic alcoholism, can lead to polyneuritis or to Wernicke's encephalopathy. Not all peripheral neuropathies, however, are due to a deficiency of thiamine. In Wernicke's encephalopathy there is an acute or sub-acute oculo-cerebral disturbance with loss of consciousness, nystagmus, ophthalmoplegia and sometimes ataxia and mental symptoms.

The methods for the determination of thiamine in plasma and urine are complicated and not within the scope of most hospital laboratories. The simplest investigation of value is the measurement of plasma pyruvate after ingestion of glucose. The resulting increased pyruvate metabolism results in greater plasma concentrations in thiamine deficiency than normal because the vitamin is necessary for pyruvate oxidation. Vitamin deficiencies in general are best investigated not by the determination of the substance itself or of its metabolites in the body fluids, but by observing improvement in the clinical condition following therapy.

Vitamin B_{12} (cyanocobalamin) deficiency is responsible for pernicious anæmia and also for sub-acute combined degeneration of the spinal cord. It seems likely that there is a specific defect in cerebral metabolism in vitamin-B_{12} deficiency. Essential features in the diagnosis are histamine-fast achlorhydria (Chapter VIII). Estimations of serum B_{12} are possible, but are less helpful than the investigation of the absorption of the vitamin labelled with the radioactive ^{58}Co (see pp. 190 and 191). After an oral dose the radioactivity of the blood, of the fæces and the urine and even the liver may be measured. In pernicious anæmia and sub-acute combined degeneration, more than 90 per cent of the radioactivity passes into the fæces, whereas less than 30 per cent normally appears. A urinary excretion test is being increasingly used.

Copper metabolism and the central nervous system. In hepatolenticular degeneration or Wilson's disease there is a congenital deficiency presumably gene-controlled of the copper-transporting protein cæruloplasmin. Copper is taken up excessively from the gut, deposited in the tissues, including the liver, kidney and brain, leading to severe neurological symptoms; it is also excreted in excessive amounts in the urine. There is also frequently a secondary amino aciduria.

The diagnosis of Wilson's disease depends upon the estimation of the level of the copper in the blood or of cæruloplasmin. This protein possesses oxidase activity with *para*-phenylenediamine as substrate, and can therefore be estimated by an enzymic method. More recent methods of investigation make use of radioactive copper, and it seems likely that it will be possible to diagnose the condition in patients before the clinical manifestations develop; prophylactic measures might then be of value.

The lipidoses. A number of clinical conditions are associated with an abnormal deposition of lipids in the tissues. Thus, there is deposition in the liver of the cerebroside kerasin in Gaucher's disease. In Niemann-Pick disease there is deposition of sphingomyelin in the brain and liver. Gargoylism is due to the deposition of an abnormal mucopolysaccharide in the tissues. Improved methods of lipid analysis have permitted recognition of some characteristic patterns of lipids deposited in the brain and Table 15 summarizes some of the recent findings in these diseases.

TABLE 15—CHARACTERISTIC FINDINGS IN VARIOUS LIPID DISEASES
(after Cumings 1965)

Disease.	Cerebral lipid abnormality.		Body fluid changes.
	Decreased.	Increased.	
Amaurotic family idiocy	Myelin lipids	Gangliosides	
Tay-Sachs	Myelin lipids	Gangliosides	
Metachromatic leucodystrophy		Sulphatide and hexosamine	Urinary sulphatide increased
Sudanophilic diffuse sclerosis	Myelin lipids	Cholesterol ester	
Globoid body diffuse sclerosis	Myelin lipids	Cerebroside	C.S.F. protein >100 mg/100 ml.
Inclusion body encephalitis		Hexosamine (Cortex and white matter)	C.S.F. protein pattern abnormal
Niemann-Pick		Sphingomyelin Gangliosides	
Gaucher		Cerebroside	
Gargoylism	Myelin lipids	Gangliosides	Urinary mucopolysaccharides increased
Alpers		Cholesterol ester (Cortex and white matter)	

Neuromuscular disorders. Familial periodic paralysis is associated with potassium deficiency as usually shown by a low serum potassium concentration. However, there has been some recent evidence that the low potassium may be secondary to movement of potassium ions into the cells. On the other hand, certain patients with an aldosterone-secreting tumour of the adrenal gland can similarly show the features of familial periodic paralysis, and these subjects without doubt suffer from potassium deficiency (see Chapter IV). The effect of deficiency of cholinesterase in the plasma on the sensitivity of the neuromuscular junction to cholinesterase inhibitors has already been considered (page 161).

Kernicterus. It is remarkable that in post-hepatic jaundice even of the greatest severity, the cerebrospinal fluid contains only traces of bile pigment and there is no staining of the brain and central nervous system. In pre-hepatic jaundice, however, the affinity for lipids of bilirubin, as opposed to that of conjugated bilirubin, leads to deposition of this pigment in brain tissue when the amount in the plasma is increased. This does not usually occur in hæmolytic jaundice in adults, because the serum bilirubin so seldom exceeds 10 mg/100 ml. However, in hæmolytic disease of the newborn and of premature infants concentrations of plasma bilirubin of 20–30 mg/100 ml. or more frequently occur, and then appreciable quantities of the pigment are deposited in the brain, particularly in the basal ganglia, which become intensely yellow, resulting in the so-called kernicterus. The severity of the jaundice occurring in hæmolytic disease of the newborn and in the premature infant is greater than is explained by the increased rate of hæmoglobin breakdown, and the functional capacity of the liver to conjugate bilirubin is known to be only about 1–2 per cent of the normal adult capacity. Jaundice may appear on the first day after birth, and plasma bilirubin may quickly increase to 20 mg/100 ml. or more. Such high concentrations may lead to kernicterus, with consequent permanent brain injury, and exchange transfusions are essential to avoid this serious complication.

Coma. The chemical pathology of coma is just as mysterious as the biochemical changes observed in sleep. Coma may be brought about by exogenous causes such as various drugs, anoxia, changes in environmental temperature and also by specific biochemical intoxication, as in uræmia, diabetic coma and hepatic coma. It can also occur secondary to cerebral damage, to toxæmias of various kinds and after epileptic attacks.

Diabetic coma is related to the accumulation of aceto-acetic acid in the blood, but as has been discussed in Chapter IX, there is little evidence that the coma is directly due to this acid. It seems much more likely that there is a disturbance of brain metabolism due to some metabolic defect which accompanies or is parallel to that responsible for the production of aceto-acetic acid. Various toxic agents retained in the blood have at one time or another been held responsible for the coma associated with uræmia. The high concentration of urea is certainly not responsible. The frequently irreversible coma of hypoglycæmia and of anoxia is obviously due to a deficiency of glucose and of oxygen supplied to the brain. The neurological disturbances associated with chronic liver disease are due to the reabsorption from the bowel of ammonia and other toxic bacterial degradation products of dietary protein (see Chapter V).

The investigation of coma due to suspected poisoning.* Confusion, unsteadiness or coma may follow poisoning by carbon monoxide, barbiturates or other hypnotics. Diagnosis may be difficult if the patient is not discovered until many hours after poisoning. Similar symptoms may arise from cerebral vascular lesions, diabetes (page 135), hypoglycæmia (page 129), uræmia or injury; these can nearly always be diagnosed or excluded by clinical examination.

If the patient is still conscious he should be asked about the type and the amount of poison suspected and the approximate time it was taken. If he is unconscious information from relatives and detective work for pills, capsules or their containers or prescriptions may be helpful, but it must not be assumed that these were necessarily the substance taken.

A catheter specimen should be examined immediately for protein, glucose, ketone bodies and salicylate. If coal gas or carbon monoxide poisoning is suspected, either from the circumstances of the incident or if the skin appears cherry-red, a blood sample should be collected for subsequent analysis for carboxyhæmoglobin. If acute poisoning by alcohol is suspected a blood sample should also be collected and preserved for subsequent analysis for alcohol by a specialist in this kind of analysis. In the absence of a strongly positive ferric chloride reaction for salicylates and in the absence of any evidence or clinical signs of carbon monoxide poisoning, it is probable that barbiturates are responsible. Gastric lavage should be performed on unconscious patients only after endotracheal intubation. The danger of washing residual fast-acting barbiturates into the small intestine,

* This is not concerned with the clinical diagnosis or treatment of suspected poisoning, but with the laboratory investigation of coma which may be due to poisoning.

where absorption is rapid, must be avoided by removing fluid from the stomach before washing out and by using small volumes of fluid for each wash (200–300 ml. for an adult patient). The stomach contents and the washouts (kept separately), as much urine as possible and 5–10 ml. of blood in citrate and C.S.F. (if the clinical condition suggests the necessity for lumbar puncture), must be put into bottles, labelled and taken personally by those handling the case to the laboratory in case they are needed for chemical analysis.

The time taken for identification or estimation of poisons limits the help which a laboratory can give in such cases. A negative test subsequently for barbiturates in stomach contents, blood or urine may indicate that the diagnosis must be reconsidered. Although an early blood analysis may be the only means of detecting fast-acting barbiturates, the widespread use of sedatives, hypnotics and tranquillizing drugs makes difficult the differentiation of therapeutic from toxic doses by estimation in stomach washings or concentrations in body fluids. Rapid quantitative measurement of barbiturates requires an ultraviolet spectrophotometer, an apparatus which should be readily available. Qualitative differentiation of short-acting barbiturates from long-acting barbiturates is also important, for low concentrations of the former are often more dangerous than the high concentrations of the latter. Such analyses may indicate the need for early treatment by forced diuresis and alkalinization or by artificial dialysis; in either case a careful watch on the fluid and electrolyte balance is important.

The laboratory investigation of poisons other than aspirin, salicylates or barbiturates is a major problem aggravated by the great pharmacological activity per unit weight of many recently developed drugs. It can be carried out only by a staff specially trained in this kind of analysis.

CHAPTER XIV

SOME BIOCHEMICAL ASPECTS OF HÆMATOLOGY

THE circulating blood of a normal adult man contains about 750 g of hæmoglobin, and of this about $\frac{1}{120}$ or about 7–8 g are degraded daily. This amount has to be newly synthesized each day because the globin part of hæmoglobin can be re-utilized only after catabolism into its constituent amino acids, and the hæm moiety is broken down into bile pigment, which is excreted; iron alone is re-utilized in the synthesis of hæmoglobin.

The rates at which hæmoglobin is synthesized and at which red cells are formed are related to the oxygen content of the blood, and therefore depend not only upon the amount of oxygen reaching the blood but also upon the capacity of the blood to carry oxygen, which in turn depends on the amount of circulating hæmoglobin. Hæmoglobin synthesis is therefore stimulated by anoxia, whether due to oxygen deficiency or to anæmia. After blood loss or hæmolysis the daily synthesis of hæmoglobin may be increased six- to eight-fold.

The details of the early stages of erythropoiesis in the bone marrow are still under discussion, but it is generally agreed that the erythrocytes are derived from primitive nucleated cells in the bone marrow by successive processes of mitosis and maturation. A primitive stem cell divides to form two cells, one of which retains its behaviour as a stem cell while the other successively divides to form two basophil normoblasts, four polychromatic normoblasts and eight orthochromatic normoblasts, after which maturation through late normoblast and reticulocyte stages to the mature non-nucleated fully hæmoglobinized erythrocyte involves no further mitotic division. These processes must involve the biosynthesis not only of hæmoglobin but also of large quantities of purine bases, nucleic acids and proteins.

The ability of the hæmopoietic tissues to manufacture erythrocytes depends on a variety of hormones, trace metals, enzymes and coenzymes and an adequate provision of essential amino acids, glycine, acetyl coenzyme A and iron. There is strong evidence that the marrow response to the stimulus of hypoxia is dependent upon a glycoprotein hormone, erythropoietin. Erythropoietin may be produced in the kidney in response to hypoxia and may act on differentiation of the stem cells rather than upon any particular step in hæmoglobin synthesis.

The role in producing disease in human subjects of deficiencies of vitamins, trace metals and co-factors known to play some part in hæmoglobin synthesis is still vague. Biotin, coenzyme A, pantothenic acid and pyridoxal phosphate are essential coenzymes required for the synthesis of hæm; of these only a deficiency of the last is known to play a role in human disease. Of the trace metals, only copper and cobalt are known to play a role, the former in the absorption of iron and the latter as an essential constituent of vitamin B_{12}. Deficiency of intrinsic factor can cause vitamin B_{12} deficiency, with abnormal maturation of red cells leading to a megaloblastic stage and consequent failure to liberate sufficient red cells to maintain a normal amount of circulating hæmoglobin, even though the average amount per cell is above normal. A similar megaloblastic anæmia occurs in folic acid deficiency.

Iron metabolism. An adult man contains a total of 4–5 g of iron distributed approximately as follows:

Hæmoglobin	60—70%
Myoglobin	3—5%
Hæm enzymes	0·2%
Plasma non-hæm iron (siderophilin)	0·1%
Storage iron (ferritin and hæmosiderin)	25%

In the adult male the daily loss of iron is small, not more than 0·5–1·5 mg, and the daily requirement is of this amount. Additional iron is required for growth during childhood and by women because of menstruation and pregnancy. So little iron is lost from the body that some regulating mechanism must operate to avoid excessive absorption and deposition of iron in the tissues; this is believed to involve the saturation of a protein, apoferritin, which combines with the absorbed iron to form ferritin. Absorption slows down when the apoferritin–ferritin is saturated; anoxia permits reduction of the iron to the ferrous state and passage into the bloodstream to combine with a specific iron-binding protein, siderophilin (or transferrin), for transport to the bone marrow for hæmoglobin synthesis, to cells for hæm protein synthesis or to the body stores. The iron complexed with siderophilin is in the ferric form, and the total iron-binding capacity is about 300–350 µg/100 ml. plasma, but of this only about a third is normally combined with iron. The plasma iron occupies a special position in hæmoglobin metabolism, for it represents a dynamic equilibrium between iron formed by hæmoglobin breakdown, iron transport from the intestine and other tissue stores and that taken up by bone marrow for hæmoglobin formation, by tissues for hæm protein synthesis, and for tissue storage. Measurements of plasma iron and the degree of saturation of the siderophilin provides information regarding iron metabolism in the body. Deficiency of iron is essentially due to blood loss with failure to replace the iron stores because of dietary deficiency, increased requirement or defective absorption. Microcytes containing a sub-normal quantity of hæmoglobin may be released into the circulation, and be ineffective in raising the hæmoglobin level to normal. Accompanying changes include brittleness of the nail and atrophy of mucous membranes.

Hæmochromatosis. Hæmochromatosis (bronzed diabetes) occurs mainly in men and is associated with a great excess of iron in the body, which is deposited in the liver, skin, pancreas, testes and other parts of the body. The plasma iron is increased and the iron-binding protein, siderophilin, is fully saturated. The cause is unknown, but the condition may be due to an abnormality of the absorption mechanism of iron or of iron transport. A secondary form of the disease is said to occur in severe hæmolytic anæmia treated with repeated blood transfusions.

Megaloblastic anaemia. This may be due to vitamin B_{12} or folic acid deficiency. Vitamin B_{12} occurs in two biologically active and interconvertible forms, cyanocobalamin and hydroxycobalamin and is absorbed in the presence of intrinsic factor (a mucoprotein) and normally stored in the liver in amounts sufficient to provide for several years.

Folic acid is pteroylglutamic acid and is reduced in the liver to tetrahydrofolic acid, which reacts enzymically with formiminoglutamic acid (FIGLU), a normal metabolite of the basic amino acid histidine. The formimino group is transferred with loss of ammonia from FIGLU to the tetrahydrofolic acid. The resulting

Folic acid   ⟶ Tetrahydrofolic Acid (THF)
Histidine   ⟶ Formiminoglutamic Acid (FIGLU)
FIGLU + THF ⟶ Formyl THF (Folinic Acid) + Glutamic Acid + NH_3

formyltetrahydrofolic acid (folinic acid) is stored in the liver and provides a mechanism for the transfer of one-carbon-atom fragments, such as formyl (CHO), hydroxymethyl (CH_2OH) and methyl (CH_3) groups, just as coenzyme A acts as a biological carrier of acetyl groups. The primary source of one-carbon-atom fragments is the amino acid serine $CH_2OH.CHNH_2.COOH$ and, in turn, glucose, from which serine is derived. Both folic acid and vitamin B_{12} seem to be concerned with the provision of such one-carbon-atom fragments for the synthesis not only of methionine, choline and other methyl compounds but also of purines and therefore of the nucleic acids so important in protein synthesis. The relationship between vitamin B_{12} and folic acid is unknown, although it has been suggested that vitamin B_{12} is essential for the formation of folinic acid and that megaloblastic erythropoiesis may be due basically to folinic acid deficiency. However, patients with pernicious anæmia are not folic-acid deficient and absorb folic acid normally.

In megaloblastic erythropoiesis there is a general disturbance of protein or nucleoprotein metabolism which leads not only to the characteristic megaloblastic bone marrow and over-hæmoglobinized short-lived red cells, poikilocytosis and anisocytosis but also to lesions of the oral, gastro-intestinal and vaginal epithelium. The megaloblastic anæmias may be due to inadequate intake, defective absorption, defective utilization or excessive demand for either or both of the factors concerned. In vitamin B_{12} deficiency the central nervous system is also affected.

Vitamin B_{12} deficiency may be investigated by the usual hæmatological investigations and assessment of gastric and intestinal function, by estimations of concentrations of vitamin B_{12} in the plasma or more conveniently by measuring the absorption of vitamin B_{12} labelled with radioactive cobalt ^{58}Co. Measurement of radioactivity of the fæces shows that 26–80 per cent of a suitable dose is normally absorbed. According to Schilling, 7–22 per cent of a dose of 1 µc ^{58}Co-cobalamin should be excreted in the urine during the following 24 hours. Other techniques involve surface counting over the liver area, total body counting or even repeated estimations of plasma radioactivity.

Folic acid deficiency may be assessed by estimating the serum folic acid, by clearance of injected folic acid from the plasma or by the estimation of urinary FIGLU after histidine loading. FIGLU may be measured by a micro-biological assay, by high-voltage electrophoresis or by chromatography, but positive results are obtained in vitamin B_{12} deficiency, in malignancy and sarcoidosis, and due to disordered histidine metabolism in cirrhosis of the liver.

The biosynthesis of hæmoglobin. The molecule of hæmoglobin consists of four hæm molecules bound to the protein globin of

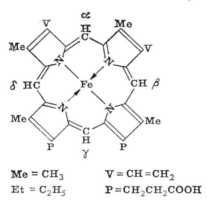

FIG. 31. Structure of hæm, the ferrous complex of protoporphyrin IX.

molecular weight about 68,000. The hæm group of hæmoglobin and of all hæm proteins is the ferrous complex of protoporphyrin IX, which bears four methyl, two vinyl and two carboxyethyl groups arranged in a special order in the β positions of its four constituent pyrrole rings (Fig. 31).

The main pathway of biosynthesis of hæm from succinyl coenzyme A and glycine is now clearly established (Fig. 32), although the details of some of the stages, especially the early ones need to be elucidated. Hæm is a highly specific and potent inhibitor of the key enzyme ALA synthetase, and this may be of physiological importance in providing a negative feedback

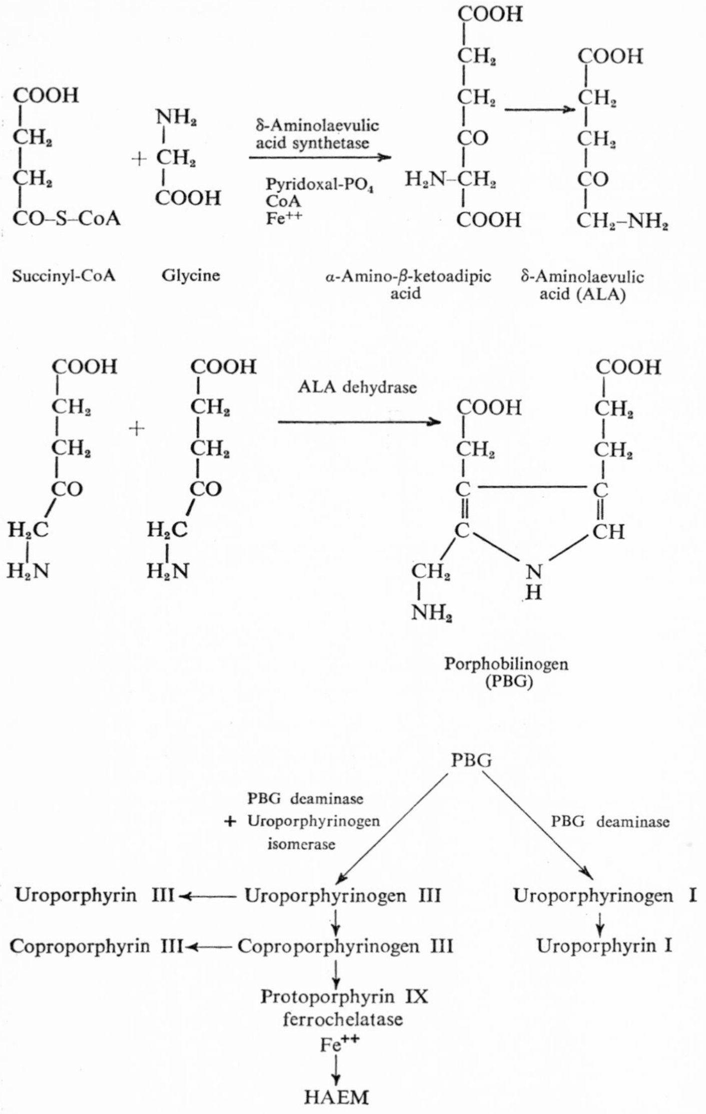

FIG. 32. Biosynthesis of hæm and uroporphyrin I.

mechanism for the control of porphyrin synthesis. If applicable to the biosynthesis of hæm in the hæmopoietic tissue, this might provide a mechanism whereby hæm synthesis can be linked to globin synthesis. When globin is present it could combine with the hæm, and only when the globin had been used up would hæm accumulate and exert its negative feedback effect on ALA synthetase and thus inhibit further porphyrin synthesis.

The globin part of hæmoglobin consists of two pairs of polypeptide chains folded complexly to give it a spheroidal shape. The detailed structures of these polypeptide chains are under genetic control and have been elucidated by Ingram and others who have determined their amino acid sequences. In normal adult hæmoglobin (HbA) there are two α chains and two β chains; the terminal amino acids of the former are valyl-leucyl while in the latter they are valyl-histidyl-leucyl residues. HbA may therefore be abbreviated to $\alpha_2\beta_2$. A change, usually genetically determined, in the amino acid sequence of either the α or β chain results in a different hæmoglobin. In fœtal hæmoglobin ($\alpha_2\gamma_2$) the β chains are replaced by γ-chains with terminal glycyl residues. In HbA_2 ($\alpha_2\delta_2$) they are replaced by still different δ-chains.

The hæmoglobins. The different hæmoglobins are due to changes in the globin moieties and may be classified into three groups. In group I there are only minor abnormalities of globin chains as in Hæmoglobin S, C, D and E. Thus, in sickle-cell hæmoglobin (see p. 161) a single glutamyl residue in the β chain has been replaced by valine. In group II the synthesis of the β chain is suppressed, and this chain is replaced by another, e.g., HbA_2 ($\alpha_2\delta_2$), HbF ($\alpha_2\gamma_2$). In group III the synthesis of the α chain is suppressed and the globin consists of four similar chains, e.g., HbH (β_4) or HbBarts (γ_4).

Abnormal hæmoglobins (only a few of which are of pathological significance) can result from mutations of the α chains (e.g., Hb I, P, Q and D) or of the β chains (S, C, D, E, G, J, L and N) of hæmoglobin A. On the other hand, some are the result of suppression of α or β chain synthesis of HbA. This is well shown by thalassæmia, of which the commonest form is due to suppression of the β chains, which results in a persistance of HbF and increased synthesis of HbA_2. Less commonly there is suppression of the α chain so that HbA_2 and HbF are also suppressed and

HbH (β_4) and HbBarts (γ_4) form. Thalassæmia minor and major occur in heterozygous and homozygous individuals respectively, and the condition is inherited as a dominant autosomal characteristic. There is an under utilization of iron, but marrow erythroid hyperplasia and anæmia lead to increased absorption of dietary iron; repeated transfusions may lead to hæmosiderosis.

Such genetic abnormalities of the synthesis of globin resulting in specific differences in amino acid composition may be reflected in an alteration in the rate of synthesis of the abnormal hæmoglobin. Thus, patients with the sickle-cell trait whose blood contains both normal and sickle-cell hæmoglobin always have less sickle hæmoglobin than normal hæmoglobin. It is possible that a genetic abnormality can result in an impairment of synthesis of one of the chains of the hæmoglobin molecule without leading to the actual formation of an abnormal hæmoglobin. There may well be rate controlling genes which act independently of genes responsible for the specific amino acid composition of chains.

Metabolism of hæmoglobin. The metabolism of hæmoglobin cannot be dissociated from the fate of erythrocytes. Apart from the methods of historical interest of assessing the dietary requirements for hæmopoiesis, the methods applicable to man include the use of differential agglutination (the Ashby technique), or ^{15}N and ^{14}C labelled glycine, ^{59}Fe after re-utilization of radioactive iron has been blocked by large doses of non-radioactive iron immediately after the administration of the radioactive material, and the ^{51}Cr labelled red cell method. Of these, the Cr method has been widely used in clinical studies, but gives only a relative measure of the red cell life, and then only if the rate of elution of isotope from the cell is constant even in disease. The ^{59}Fe method has been of limited value in man because of the problem of giving sufficient iron to block re-utilization of radioactive iron. The use of glycine labelled ^{15}N or ^{14}C is probably the most satisfactory method, but ^{15}N requires an expensive mass spectrometer, whereas ^{14}C requires highly sensitive equipment for measuring low-level radioactivity.

In all the species so far studied the normal red cell has a finite life span. There is in addition some random destruction which varies from species to species. No pathological condition has yet been found in which there is a lengthening of the life span, but in a number of conditions there is a shortening which may be due to premature senescence or may occur as a result of a random process affecting the cells independently of their age. There has been much effort to determine the mechanisms which limit hæmolysis in the various forms of hæmolytic anæmia. With the possible exception of very recently formed erythrocytes, the younger cells are much more resistant to destruction than the older cells. One can therefore understand if the action of a hæmolytic agent is selective in affecting mainly older cells, such resistance must be important in limiting hæmolysis.

The integrity of the red cells depends upon the continuous supply of energy in the form of adenosine triphosphate (ATP) to maintain the characteristic ionic pattern, including the high concentration of potassium within the red cell; the red cell

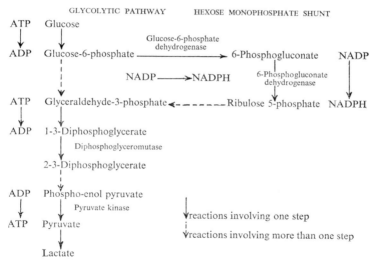

FIG. 33. Role of enzymes of carbohydrate metabolism, deficiency of which can be associated with abnormalities of the red cell.

membrane is dependent upon maintenance of glutathione in the reduced form by a continuous supply of NADPH (TPNH, see p. 131), which together with NADH (DPNH) is required for the reduction of methæmoglobin continuously formed by the oxidation of hæmoglobin. The normal red cell reaches the end of its life span when the supply runs out of precursors and intermediates necessary for the supply of ATP and NADPH. ATP is formed by the glycolytic pathway and NADPH by the hexose monophosphate shunt, which provides about 10 per cent of the energy required by the red cell. Fig. 33 shows some of the sites in the metabolism of glucose at which enzyme deficiency has been shown to be responsible for increased hæmolysis.

Hæmolytic anæmias. Hæmolysis of red cells may occur before the end of a normal life span because of inherited enzyme deficiencies which may be sufficiently severe in themselves to produce hæmolysis, or which if insufficiently severe may become so when the subject is exposed to an additional toxic agent.

Phenylhydrazine is a potent hæmolytic agent and causes hæmolysis in all subjects, but other drugs, such as sulphonamides and sulphones, are more variable in their effect and only cause hæmolysis in a few sensitive subjects. The enzyme abnormality responsible for these is not known, but the sensitivity to primaquine, an antimalarial, shown by many negroes, is known to be related to a glucose-6-phosphate dehydrogenase deficiency due to a sex-linked gene of intermediate penetrance. The red cells of such individuals form Heinz bodies when exposed *in vitro* to phenylhydrazine, and may be shown to contain low concentrations of reduced glutathione and of glucose-6-phosphate dehydrogenase. This may be estimated by accurate enzyme analysis, but useful screening tests are provided by a rough estimation of the extent of the enzyme deficiency using bromcresyl blue or by an assessment of glutathione stability towards acetylphenylhydrazine.

Glucose-6-phosphate dehydrogenase deficiency of red cells is not uncommon in the Mediterranean races and in those of Caucasian descent, and accounts for the high incidence of hæmolytic disease of the newborn in Greece and for the incidence of favism among Mediterranean races. In this last condition the enzyme deficiency leads to sensitivity to broad beans, and in subjects in whom the gene has been fully expressed, exposure even to the pollen may cause a severe hæmolytic anæmia.

Congenital non-spherocytic hæmolytic anæmia is usually due to a deficiency of pyruvate kinase, but may also occur in association with phosphoglyceromutase or even glucose-6-phosphate dehydrogenase deficiency. Autohæmolysis of incubated red cells of subjects with this condition may be decreased by the presence either of glucose or of ATP according to the enzyme defect.

Hereditary spherocytosis is also accompanied by a failure of glucose to decrease the rate of hæmolysis of the red cells. In this case there seems to be decreased formation of ATP, which might either affect the ionic composition within the red cell, and cause hæmolysis, or perhaps interfere with lipid synthesis and cause abnormality of the red cell membrane. Splenectomy is said to be curative by removing a large number of metabolizing cells which compete for glucose as a substrate and cause destruction of susceptible cells.

Paroxysmal nocturnal hæmoglobinuria. In this condition the red cell is abnormally sensitive to properdin, complement and magnesium ions, which form a lytic system with an optimum activity at pH 6·7–7·1. The red cell is said to appear abnormal and to be deficient in cholinesterase. Such subjects might therefore become more acidotic than normal during sleep, and hæmolysis would occur. Diagnosis depends upon the demonstration of hæmolysis greater than normal when the red cells are placed in a buffer at pH 6·8 in the presence of complement (Ham's test).

Auto-immune hæmolytic anæmia. Auto-antibodies may be warm or cold according to whether they react at 37° or 20°. The former are γ-globins of low molecular weight and are not hæmolytic *in vitro*, but damage cells *in vivo* so that they become sequestered in the spleen. The cold auto-antibodies are γ- or β-cryo- or macro-globulins and may be hæmolytic *in vitro*. Cold auto-antibodies may be harmless unless their titre and residual activity at 37° becomes great enough to cause hæmolytic anæmia. In paroxysmal cold hæmoglobinuria Donath–Landsteiner antibodies are present. These are responsible for cold–warm lysis, i.e., they are absorbed to cells at 20° and then cause lysis in the presence of complement at 37°. 65 per cent of patients with auto-immune hæmolytic anæmia are primary and associated with auto-antibodies; the remaining 35 per cent are secondary to other diseases, such as reticulosis, leukæmia, reticular-cell sarcoma, disseminated lupus erythematosus, rheumatoid arthritis or ulcerative colitis.

Other acquired hæmolytic anæmias are associated with renal disease, tuberculosis, sub-acute bacterial endocarditis, malignant disease, microangiopathic hæmolytic anæmia, leukæmia, March hæmoglobinuria (due to mechanical damage), extra-corporal circuits and teflon grafts.

Bile-pigment formation from hæmoglobin. The administration of isotopically labelled glycine to human subjects for the investigation of life span of red cells is followed by the excretion of labelled bile pigment 80–140 days later. This labelled bile pigment is formed by the breakdown of the labelled hæmoglobin in red cells at the end of their life span. There is also an early peak of excretion of labelled bile pigment which cannot be thus accounted for. In normal subjects this early labelled fraction of bile pigment amounts to about 5–10 per cent of the total excretion of bile pigment, but there is a much higher proportion of this fraction in the bile pigments in pernicious anæmia, sickle-cell anæmia, thalassæmia, aplastic anæmia and a number of other conditions not necessarily accompanied by anæmia, for example, congenital porphyria and one form of congenital non-hæmolytic pre-hepatic jaundice.

Thrombosis and fibrinolysis. Coagulation of blood and fibrinolysis of the clot probably take place simultaneously throughout life and become of pathological importance only when the balance of the two phenomena is disturbed by disease leading to thrombosis or systemic fibrinolysis.

A simple cascade sequence is responsible for the formation of the various protein factors concerned with the production of a fibrin clot. Each protein factor occurs in plasma in an inactive or precursor form which reacts, in a stepwise sequence, with a specific substrate to form another active enzyme. Interaction eventually converts prothrombin to thrombin, which then converts fibrinogen to fibrin. These reactions, summarized in Fig. 34, have been recognized by use of plasma from patients with congenital deficiencies and by partially purified preparations of clotting factors.

The fibrin of a clot is hydrolysed into soluble polypeptides by a proteolytic enzyme, plasmin. This is formed from its precursor plasminogen, a normal plasma β-globulin, by the splitting off of lysine–arginine bonds, by the action of activators which occur in high concentration in various tissues, urine and other body fluids. The concentration of activators is increased after exercise, adrenalin or pyrogens, and activator may be produced from the walls of the veins by local anoxia. The proteolytic activity of plasmin is antagonized by anti-plasmin occurring in the plasma and in the platelets. When thrombus formation occurs adsorption of activator to the fibrin results in local plasmin formation and thrombolysis may follow.

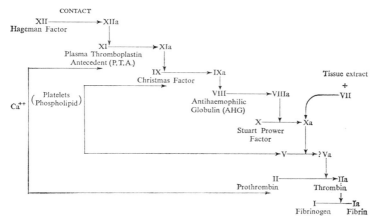

FIG. 34. The cascade mechanism in blood coagulation according to Macfarlane. The processes concerned with factors XII to VIII are relatively slow and are completed in minutes; the remainder of the processes are very fast and require only seconds for completion. (Measurement of Prothrombin Time involves the reactions from factor VII to the formation of fibrin.)

Systemic fibrinolytic states are common after thoracic surgery, the use of heart–lung machines and obstetric hæmorrhage, and are due to the sudden release into the circulation of activators of plasminogen in sufficient concentration to overwhelm the natural inhibitors. In these conditions there is presumably a preliminary release of thromboplastin to the circulation, with excessive intravascular clotting followed rapidly by an excessive physiological response of increased fibrinolytic activity. In therapy, streptokinase, an activator of plasminogen prepared from hæmolytic streptococci, and urokinase, an activator present in normal urine, are used as plasminogen activators in the treatment of thrombosis, while ε-aminocaproic acid is a potent inhibitor of plasminogen activator and may be of value in the treatment of systemic fibrinolysis.

Further Reading

HUISMAN, T. H. J. 1963. "Normal and Abnormal Human Hæmoglobins." Advances in Clinical Chemistry. Eds. H. Sobotka and C. P. Stewart. Vol. 6. Academic Press. p. 232.

PRANKERD, T. A. J. 1965. "Clinical Significance of Red Cell Structure and Metabolism." *Brit. med. J.* **2**, 1017.

"Symposium on Disorders of the Red Cell." *Amer. J. Med.* 1966, **41**, 657.

"Symposium on the Mechanisms of Disorders of Erythropoiesis." *Medicine,* 1964, **43**, 625.

CHAPTER XV

SOME MISCELLANEOUS TOPICS

Nutritional deficiencies. Apart from starvation, with consequent loss of weight, conditions are sometimes encountered in which only one or some of the dietary constituents are deficient. Protein deficiency due to ignorance or poverty is probably the most common. There is a diminution of all the fractions of the serum proteins which is readily shown either by the classical determination of serum albumin and globulin or, better, by the paper electrophoresis technique. Œdema may be present, but is not directly related to the hypoproteinæmia. Deficiency of fat in steatorrhœa, whether due to the sprue syndrome or to pancreatic disease, may lead to deficiency of vitamins A and D; indeed, a low concentration of vitamin A in plasma is of value in the diagnosis of congenital fibrocystic disease of the pancreas.

Vitamin deficiencies. It is possible to measure concentration of most vitamins in body-fluids, but many of the techniques require facilities available only in laboratories specializing in this field. Only the determination of vitamin C is within the scope of an ordinary hospital laboratory.

Vitamin A deficiency may occur because of dietary deficiency or because of an impairment of fat absorption, and may be demonstrated by determination of the vitamin and its precursor, β-carotene in plasma.

Ascorbic acid (vitamin C) deficiency may present itself as a frank and obvious scurvy or as a sub-clinical purpuric condition more difficult to diagnose. This vitamin is readily determined in urine and also with somewhat greater difficulty in the blood plasma and leucocytes. A single determination of ascorbic acid in urine is useless, but the increase in urinary excretion after large daily doses of the vitamin (see p. 217) is by far the best and simplest test available. However, the absence of vitamin C deficiency can more rapidly be established if the vitamin can be detected in the plasma. It is claimed that the concentration of ascorbic acid in the leucocytes provides an excellent indication of the state of nutrition in respect of this vitamin; severe deficiency with symptoms is associated only with negligible amounts of leucocyte ascorbic acid.

The investigation of thiamine (vitamin B_1) and of vitamin B_{12} deficiency have been discussed in Chapter XIII.

Deficiency of nicotinamide, the anti-pellagra vitamin, may readily be determined from the decreased elimination in the urine of the end product of its metabolism, N-methyl nicotinamide. The rarity of the condition in Britain prevents most laboratories from acquiring necessary experience in the determination.

Vitamin deficiencies are best investigated not by the determination of the substance or its metabolites in body fluids, but by observing improvement in the clinical condition following therapy. This is well shown in vitamin D deficiency, in which characteristic changes in calcium, phosphorus and phosphatase of the serum are present. There is no satisfactory method for the determination of vitamin D, but therapeutic treatment with the vitamin should rapidly and specifically correct the clinical condition and restore the blood chemistry to normal.

Transport of fat. It is customary to regard abnormalities of transport of fat to be revealed by a high fat content in the plasma, but it must be remembered that this may be due either to an excessive passage into the bloodstream or to a delayed removal of fat and that there may in fact be no true abnormality of transport. Of the various fats in blood, cholesterol is the most readily, though not very accurately, determined, and the clinical conditions which are associated with changes in the concentration of this compound are considered on p. 205. Circulating cholesterol and other lipids are transported in the plasma in association with protein and form macro-molecular lipid–protein complexes. Cholesterol is the more readily investigated of the blood lipids, and it has been shown that 60–70 per cent of the plasma cholesterol is associated with the α-globulin, while 30–40 per cent is associated with β-globulin. This has been shown by electrophoresis, either in the classical apparatus of Tiselius or on filter-paper, or by ultracentrifugation. In the last technique plasma is diluted with suitable media of various densities and centrifuged at high speed. The macro-molecules then sediment according to their specific gravity. The β-lipoproteins are found to have low density and the α-lipoproteins to have a high density. Patients with coronary disease are found to have a diminished amount of α-lipoproteins and an increase in β- or low-density proteins. Recently paper electrophoresis has been alleged to show a specific free β-lipoprotein characteristic of nephrosis, diabetes and after coronary thrombosis, but tests used in paper electrophoresis are not very specific, and the stains used for staining fat are not pure and can give spurious results.

Fats and atheroma and coronary disease. There is increasing evidence that myocardial infarction is associated with a chemical disorder of transport of blood fat which may lead to atheroma of the coronary arteries as well as to an increased coagulability of the blood. Duguid has emphasized the view that atheromatous lesions are secondary to minute thrombi and that these may be due to a primarily chemical disorder of the transport of blood fats. It is characteristic that atheroma is particularly common in obesity, diabetes, myxœdema, nephrosis and idiopathic lipidosis, which are all associated with high levels of plasma cholesterol and other fats. However, the mean level of the plasma cholesterol is independent of the dietary content, for it is mostly synthesized from acetate. Nevertheless, comparison between the diets of dwellers in Minneapolis and Naples and between Bantus and Europeans have shown that Neapolitans and the Bantus, whose diet is characteristically low in animal fats, are less liable to get coronary disease, and this has led some workers to believe that it is the total fat content rather than the cholesterol content which is concerned with the concentration of cholesterol in the plasma. There is no doubt that patients with coronary disease have high plasma cholesterol levels for their age, and Keys believes that a high fat diet leads to a high plasma cholesterol, which can lead to atheroma and the development of coronary disease. The evidence is based on statistical analysis, and there is great individual variation; it is therefore clear that other factors are concerned. For example, onset of atheromatous change in coronary disease in women is alleged to be deferred by the circulation of œstrogens during fertile life. At present it seems that there is no evidence that lowering the serum cholesterol by administration of diets low in saturated fats and high in unsaturated fats is necessarily associated with lowered incidence of coronary disease.

Goffmann's studies have suggested that there is an abnormality of the physical state of fat in blood. He has shown that the so-called S10 to S20 lipoprotein particles are increased in patients who not only have coronary disease but who subsequently develop the condition. Other workers believe that so-called fat-clearing factors are concerned. Heparin can clear the plasma opalescence which occurs after a fatty meal. It is also known that the capacity of heparin to clear that opalescence after a fatty meal is diminished in subjects suffering from coronary infarction as compared with controls of the same age and sex. Thrombin formation is undoubtedly related to certain of the phospholipids. Thus, it is known that lecithins, especially ethanolamine phosphatide, can reduce clotting time. Fat absorption is alleged to be accompanied by an increase in ethanolamine phosphatide which results in an increased tendency to clotting.

Abnormalities of fat deposition. These include the lipodystrophies, of which the most common is the local form occurring after repeated insulin injections; a generalized form, especially of the lower part of the body, is a well-recognized, though rare, syndrome. Localized fat depositions occur in the form of lipomata either solitary or multiple, as well as in adiposis dolorosa. Finally, abnormal deposition of fat must be considered in relation to obesity, which is essentially due to an excess of dietary intake of calories above those required. Thus it may be due to over-eating, to lack of exercise, a familial form probably due to a familial propensity for over-eating, or to laziness. Endocrine disorders of obesity include the characteristic female distribution of fat, myxœdema, Cushing's disease, and adipose gynism and gynandrism in childhood. In the lipidoses there is abnormal deposition of lipid in certain tissues. (See p. 183.)

Plasma or serum uric acid. The determination of the serum or plasma uric acid concentration by most of the customary methods is unspecific and far from satisfactory. Interpretation of the result obtained must take into account the concentration in normal persons by the method used. The usually accepted range is 2–5 mg/100 ml. in females and 4–7 mg/100 ml. in males. In acute gout 10 mg/100 ml. is sometimes observed. Figures even higher are regularly obtained in severe uræmia and often in leukæmia. In chronic gout the uric acid concentration is slightly raised, and it is a useful diagnostic test to look for the fall to normal values sometimes resulting from treatment with colchicine.

Plasma or serum cholesterol. The determination of serum or plasma cholesterol is also highly unsatisfactory, and during fasting only figures above 300 mg/100 ml. should be regarded as increased. High figures are obtained in myxœdema, cretinism, the nephrotic syndrome, jaundice, diabetes, pregnancy, some cases of arteriosclerosis and hypertension and some forms of xanthomatosis. The low values demonstrated by statistical analysis in hyperthyroidism are of no diagnostic significance in the individual patient.

APPENDIX

Collection of blood. The technique of collection of blood is best learned at the bedside from an experienced teacher but the following points are worthy of special mention:

1. The needle should have a short bevel.
2. Blood should be collected with minimum of stasis.
3. The vein used is better felt than seen.
4. If the veins look difficult the arm should be compressed with a sphygmomanometer cuff inflated to just below the diastolic blood pressure.
5. Before removing the needle from the vein the cuff should be let down and, if possible, loosened.
6. The quantity of blood collected must be correct for the amount of anticoagulant used.
7. The blood must be gently mixed with the anticoagulant.
8. Whole blood collected with a suitable anticoagulant may be used for the analysis of any constituent distributed equally or about equally between red cells and plasma (e.g., urea and glucose).
9. Loss of blood glucose by glycolysis should be prevented by the addition of fluoride, but this must not be used for specimens for urea analysis.
10. *Serum or plasma* are required for all analyses of substances distributed unequally between cells and plasma (e.g., sodium, potassium, bilirubin, phosphates, etc.). The anticoagulant used to obtain plasma must not contain the substance for which the material is analysed.
11. *Serum* must be used for calcium determinations, because most of the common anticoagulants form an insoluble or non-ionized calcium salt which would interfere with the estimation.
12. *Plasma* must be used if the substance estimated is likely to pass out rapidly from the cells during separation (e.g., phosphorus would be liberated from the organic phosphates of the red cells if they were not quickly separated—within 15 minutes of collection). Serum cannot usually be

obtained in reasonable yield within 1 or 2 hours of collection.
13. *True plasma* must be used for Cl and HCO_3 determinations.
14. Serum for enzyme determinations should be separated within 1–2 hours and kept in the deep freeze ($-20°$) until analysed.

ROUTINE TESTS

URINE

Protein. In reliability, sensitivity and simplicity there is little to choose between the tests to be described, although the last two have the advantage of not requiring the application of heat.

(*a*) *Boiling test.* Fill a test-tube three parts full with clear urine (filter if necessary), incline at an angle and boil the upper layer by means of a small flame. A turbidity indicates either protein or phosphates. Add one drop of strong acetic acid and boil again. Any remaining turbidity indicates the presence of albumin. Sometimes the turbidity appears only after the addition of the acetic acid.

(*b*) *Salicyl-sulphonic acid test.* Place 5 ml. of urine in a test-tube and filter if cloudy. Add 1 ml. of 25 per cent salicyl sulphonic acid. If protein is present the urine in the test-tube presents a turbid white appearance.

(*c*) *Albustix.* This is a strip test for protein which is extremely sensitive and is able to detect as little as 30 mg or less per 100 ml. of urine. A small strip of absorbent paper is impregnated with an indicator of appropriate range. On dipping the end in the urine free from protein the yellow colour of the strip is unchanged. In the presence of 30 mg or more of protein per 100 ml. the protein is absorbed upon the filter-paper and produces the well-known protein effect on the range of the indicator so that the paper changes from yellow through various shades of green to blue, which indicates a large amount of protein, probably 1 g or more per 100 ml. urine.

Urine should not be tested that has been collected by catheter or into a bed bottle disinfected with cetavlon, hibitane or chlorhexidine. Such antiseptics give a blue-purple colour with Albustix.

Bence-Jones protein. This is often present in multiple myeloma and has been discussed on pp. 90–92. The following test, due to Bradshaw, is a useful preliminary. A few ml. of urine are superimposed on an equal volume of concentrated hydrochloric acid. Bence-Jones' protein gives a heavy curdy precipitate at the junction of the two fluids. If the test is negative no further investigation is necessary, but when it is positive the urine must be further examined. If the urine is alkaline to litmus, neutralize, using 33 per cent acetic acid. Place a thermometer and a test-tube containing about 5 ml. of urine in a beaker of water. Heat cautiously while stirring, and observe the temperature at which the precipitate appears. When Bence-Jones' protein is present the urine suddenly becomes turbid between 40° and 60° C, but clears again partially or completely on boiling. Not all specimens behave typically, and special investigations are sometimes necessary.

Reducing substances ("sugar"). Reducing substances may be detected by Benedict's test, or by the Clinitest tablet test. Benedict's and the standard Clinitest are of about equal sensitivity and permit an estimation of the concentration of glucose in the urine, but the former requires great accuracy in measuring the urine and reagent, as well as special care in applying heat; Clinitest requires no external source of heat. Both techniques detect reducing substances and are subject to similar interference by non-glucose-reducing substances. In contrast, the enzyme tests (see below) are specific for glucose, but cannot give any indication of the concentration of glucose present.

In the routine investigation of adult subjects one requires a screening test capable of detecting concentrations of 0·1 g/100 ml. or more of glucose in urine. The enzyme test will do this very well and a negative result is unequivocal. It is useful, however, to check a positive result by assessing roughly the concentration of glucose in a patient's urine with Clinitest.

In infancy and childhood the use of the enzyme test as a routine screening test might lead to a failure to recognize congenital galactosæmia (see page 167), for the adequate treatment of which early diagnosis is essential. Although the condition is very rare,

Benedict's test or Clinitest should be used in testing the urine of children. In older children, say above the age of ten years, and in adults, the use of the simpler enzyme test for screening purposes is adequate, because failure to detect other rare abnormalities, such as pentose or fructose, is not likely to have such serious effects.

In the control of treatment in known diabetics the presence of small amounts of glucose is of little importance, but it is important to distinguish with reasonable certainty between 0·5 and 2 g or more per 100 ml. Of the tests described below, only the Benedict's test and Clinitest can do this.

Benedict's test. To 5 ml. (1 in. in a 6 in. × $\frac{5}{8}$-in. test-tube) of Benedict's reagent add eight drops of the urine. Boil vigorously for 2 min., and (if necessary) allow to cool spontaneously. If glucose is present the entire body of the solution will be filled with a precipitate, which will be red, yellow or green in colour, depending on the amount of sugar.

Note. It is essential to add the correct volume of urine. If too much be added the results will be ambiguous.

Clinitest. With the special dropper in an upright position, place 5 drops of urine in a test-tube. Rinse the dropper and add 10 drops of water. Drop in a Clinitest tablet and watch while the reaction takes place. Do not shake the test-tube during the reaction nor for 15 seconds after boiling has stopped. After 15 seconds' waiting period, shake test-tube gently and compare with the colour scale.

Negative: no sugar—the fluid will be blue at the end of a waiting period of 15 seconds. All shades of blue are negative. The whitish sediment that may form has no bearing on the test.

Positive: sugar present—the fluid will change colour. The more sugar, the greater the change and the more rapidly it occurs. The amount of sugar is determined by comparing the test-tube with colour chart at the end of the 15 seconds' waiting period.

Enzyme test for glucose

Clinistix. One end of a specially treated thick paper strip, little larger than a book match, is dipped into the urine and withdrawn. The end of the strip turns blue within 1 minute if glucose is present; if the stick is still colourless at the end of this time there is no significant glucose in the specimen.

The strips are impregnated with glucose oxidase, *o*-tolidine and a vegetable peroxidase. When the strip is dipped in urine containing glucose and withdrawn the glucose in the minute amount of urine absorbed on the strip is oxidized by atmospheric oxygen in the presence of the oxidase. Gluconic acid and hydrogen peroxide are formed, and this last product subsequently reacts with *o*-tolidine in the presence of the peroxidase to produce a blue or green colour.

The reaction is inhibited by very high concentrations of ascorbic acid, such as may occur in urine after the ingestion of therapeutic doses of the vitamin and after the injection of certain tetracycline preparations which are stabilized with ascorbic acid.

Ketone bodies

Rothera's test for acetone and aceto-acetic acid. To half a test-tube of urine add 2 cm of the mixed crystals of ammonium sulphate and sodium nitroprusside (powdered in proportion of twenty parts of sulphate to one part of nitroprusside). Then add 1 ml of strong ammonia, place the thumb over the mouth of the test-tube, shake well and allow to stand. A characteristic permanganate coloration indicates the presence of acetone, or aceto-acetic acid, or both.

The test is much more sensitive for aceto-acetic acid than for acetone. The reaction with acetone in undiluted urine can just be determined in a dilution of one in 10,000, while aceto-acetic acid shows it in a one in 125,000 dilution. The intensity of colour and rate of development vary with the concentration.

Gerhardt's test for aceto-acetic acid. To 1 in. of the urine in a test-tube add 10 per cent ferric chloride solution, drop by drop, until no further precipitate of ferric phosphate is formed. Filter. To the filtrate add some more ferric chloride solution. A Bordeaux red colour indicates aceto-acetic acid. The reaction is not obvious in a dilution of more than one in 1,500.

Note. A similar colour, not, however, bleached by adding glacial acetic acid, is given by a large number of substances such as salicylic acid, and the compounds excreted after the administration of aspirin, antipyrin, etc.

Boiling converts aceto-acetic acid into acetone. If urine which has been boiled for some time colours with ferric chloride solution the colour is not due to aceto-acetic acid.

Acetest. These tablets are used by nearly all authorities in the diagnosis and treatment of diabetic coma. The test is less sensitive than the Rothera test, which is capable of detecting minor degrees of ketosis of no clinical significance. It is more sensitive than the ferric chloride (Gerhardt) test, but is not subject to interference by salicylates.

Place an Acetest tablet on a clean white surface, and put 1 drop of urine on to the tablet. Compare the tablet with the colour scale at 30 seconds. Report as negative, trace, moderate, strongly positive. A strip test based on the nitroprusside reaction is now available.

Bile pigments

Bilirubin

The iodine and nitric acid tests are insensitive and should no longer be employed. Harrison's and the Ictotest tablet tests are reliable and sensitive, but the latter has the advantage that it requires no filtration. A test, less sensitive than Harrison's, but just as sensitive as the out-of-date iodine test, consists of shaking the urine. A yellow foam indicates bilirubin.

Harrison's test. To about 10 ml. of urine in a test-tube add about 5 ml. of 10 per cent barium chloride. Mix and filter. After the fluid has passed through, the paper is spread on another piece of filter-paper and one or two drops of Fouchet's reagent (25 g trichloracetic acid, 100 ml. H_2O, 10 ml. 10 per cent ferric chloride) are added to the precipitate. A green or blue colour indicates bilirubin.

Ictotest. Place 5 drops of urine on 1 square of special "Ictotest" mat (either side may be used). Put an "Ictotest" tablet on the mat and allow 2 drops of water to flow on the tablet. Read within 30 seconds.

Positive: the mat around the tablet turns bluish purple. (Disregard the colour of the tablet.) The amount of bilirubin is proportionate to the speed and intensity of colour.

Negative: the mat shows no colour within 30 seconds or may appear pink.

Urobilin and urobilinogen

Ehrlich's test for urobilinogen. Add 1 ml. of Ehrlich's aldehyde reagent to 10 ml. of fresh undiluted urine and allow to stand 3 minutes; red colour suggests urobilinogen is present.

Schlesinger's test for urobilin and urobilinogen. To about 10 ml. of urine in a test-tube add 6 drops of tincture of iodine (to oxidize any urobilinogen into urobilin). In another tube place 10 ml. of absolute alcohol and about 1 g of powdered zinc acetate. Pour the contents of one tube into the other and then back into the empty tube. Repeat the process until the zinc acetate has mostly gone into solution. Filter and examine by transmitted and by reflected light. A fluorescence is due to a compound of zinc with urobilin. Spectroscopically there is a characteristic absorption band at the junction of the green and blue.

When the Schlesinger test is negative, or only weakly positive with a urine giving a strong Ehrlich's reaction, the presence of porphobilinogen should be suspected and special investigations undertaken, for acute porphyria may be the condition present.

Chlorides. To 2 ml. of the urine in test-tube washed out with distilled water, add a few drops of nitric acid and 2 ml. of a 3 per cent solution of silver nitrate. An abundant curdy precipitate of silver chloride appears at once. If the chlorides are less in quantity the solution merely appears milky or opalescent.

Note. If nitric acid is not added the urates might be precipitated, especially if the urine be ammoniacal. If protein is present in the urine it must first be removed as follows:

Heat half a test-tube of urine in a water-bath. Add a piece of litmus paper and acetic acid till just acid. Boil again. Filter. Test filtrate for chlorides.

Blood. For many years microscopy has been held to be the most satisfactory test for blood in urine, but the results of microscopy are often invalidated by hæmolysis in the urine before or after voiding. Moreover, the presence of numerous pus cells may make the recognition of red cells difficult. The advantages of a chemical test for screening purposes, such as during anticoagulant therapy or following known diseases of the kidney or urinary tract, or bleeding disease, are therefore obvious. The chemical method and the examination of the centrifuged deposit have been shown to be of about equal sensitivity. The first, however, has the advantage of ease and speed of performance and the second the merit of allowing the observation of other formed elements in the urine.

"Occultest" for blood in urine. Place 1 drop of mixed uncentrifuged urine on a test paper. Place an Occultest tablet in the centre of the moist area of the paper. Allow 1 drop of water to flow on the tablet, wait 5–10 seconds and flow a second drop on the tablet so that it runs down the sides on to the test paper.

Negative: If within 2 minutes a blue colour does not appear on the test paper around the tablet the test is negative.

Positive: An area of the test paper turns blue. The amount of blood is proportional to the time of appearance and to the intensity of colour. (Colour on the tablet is of no significance.)

Porphobilinogen. In addition to circumstances in which porphyria is suspected, this investigation should be carried out in all surgical emergencies, nervous diseases and psychiatric conditions in which the diagnosis is obscure. As indicated above, a positive Ehrlich's reaction in the presence of a negative Schlesinger test may indicate the presence of porphobilinogen. The following confirmatory test, due to Rimington, should be carried out:

To 1 ml. of fresh urine add 1 ml. of Ehrlich's solution. After 1·5 min. add 2 ml. saturated sodium acetate solution and 2 ml. amyl alcohol–benzyl alcohol mixture (3:1 (v/v)) and shake. After separating, a red colour in the upper layer denotes urobilinogen, but a red or pink colour in the lower layer denotes porphobilinogen.

Phenylketonuria (see pp. 163–165). This test should be carried out as a routine on all babies between three and four weeks and again between six and eight weeks of age.

Three drops of 10 per cent aqueous ferric chloride are added to 1 ml. of urine. A cloudy or clear grey-green or blue-green colour indicates the presence of phenylpyruvic acid. The test may be carried out on the wet napkin.

Phenistix is now widely used for the screening of infants. Its advantages over the ferric chloride test are that it is simpler to use, more specific and, being unaffected by changes in temperature, pH and phosphate content of urine, has a greater overall sensitivity than ferric chloride. The stick should be moistened by dipping momentarily into a sample of urine or by pressing between the folds of a freshly wetted napkin. A positive result is shown by the appearance of a grey-green colour at the tip within 30 seconds. The colour scale provided is not truly quantitative.

Special tests for the estimation of blood phenylalanine should be carried out for confirmation of the diagnosis and are also now being widely used in screening procedures.

FÆCES

o-Tolidine test. A thin emulsion of the fæces is prepared, with the addition of water if necessary, and a smear of the suspension is made with an orange stick on to a strip of Whatman No. 121 drop reaction paper. The paper is turned over and one drop of 1·2 per cent *o*-tolidine in glacial acetic acid is dropped on. One drop of 10 vol. per cent of hydrogen peroxide is added to the reverse side of the paper. A positive test is indicated by a green to blue colour developing within 30 seconds.

"Hematest" test for occult blood in fæces. Place specimen of fæces (small smear) on filter-paper. Place one Hematest tablet in centre of specimen and flow 1 drop of water on tablet; after a period of 5–10 seconds flow a second drop of water on tablet to run down on to the paper.

Negative: If within 2 min. a blue colour does not appear on the paper the test is considered negative.

Positive: The moist area of filter-paper turns blue. The amount of blood is proportional to the colour intensity and the time it first appears. (The colour on tablet is of no significance.)

Hematest is less sensitive than the *o*-tolidine test as performed above and may be carried out without any preliminary dietary régime on fæces or even on the finger-stall used for rectal examination. "Occultest", a test designed specifically for the detection of hæmoglobin in urine, is more sensitive, and if used for the detection of occult blood in fæces demands preliminary preparation of the patient in just the same way as is necessary in the benzidine test.

Hemastix. An orange stick is thrust into the stool and a minute amount of the adherent fæces smeared over a freshly moistened hemastix placed on a white tile or other suitable surface. Alternatively, the moistened test strip may be laid on the stool on a freshly exposed surface cut by a disposable wooden spatula. Care must be taken not to moisten the hemastix too thoroughly or the reagents will dissolve out. A positive test is indicated by a blue colour appearing either within 15 seconds or within 30 seconds according to the sensitivity required of the test (see pp. 124 and 125).

It has been customary to test a single small piece of fæces and to assume that the sample is representative of the whole stool. The assumption that stools are homogeneous is unwarranted, and uneven distribution can occur with hæmoglobin breakdown products, particularly when bleeding started and ceased abruptly. Tests should be carried out on three different parts of the same stool, and only if these do not agree should further samples be examined. Since stools collected on three consecutive days need to be examined, the simpler tests are to be preferred to avoid prolonging an unpleasant job.

SPECIAL TESTS

The investigation of proteinuria

1. *Without physical signs or symptoms.* (*a*) Microscopic examination for white blood cells, red blood cells (see chemical test), casts (hyaline, blood, granular, fatty and amyloid in urine).

(*b*) Test early morning specimen passed while the patient is still lying down. If protein present, carry out renal function tests. If absent, the case is one of orthostatic albuminuria.

2. *With physical signs.* The following renal function tests should be performed:

(*a*) *Blood urea.* If there is any possibility of uræmia, determine the blood urea. If raised, no other tests are usually necessary, although sometimes measurement of the specific gravity of the urine may enable differentiation of pre-renal uræmia from renal uræmia; in the former, specific gravity may be high, while in the latter it may be fixed at 1·008–1·012. The blood urea should be carried out at appropriate intervals to follow the progress of the case. If the blood urea is normal, then the other tests of renal function must be carried out.

(*Note.* A high blood urea may be pre-renal, renal or post-renal. The first may be due to cardiac failure, alkalosis or salt deficiency, and the last to obstruction of the urinary tract.)

(*b*) *Specific gravity tests.* Determine the specific gravity of the urine 1 hour after drinking 1 litre of water and again after restricting fluid for 12 or 24 hours. The specific gravity should increase from 1·002 to 1·030 approx.

(c) *Urea concentration test.* This is no longer used.

(d) *Urea clearance test.* The clearance is the maximum quantity of blood the urea content of which is completely removed per min. by passage through the kidneys. It may be determined at any time providing there is no sudden diuresis due to intake of excess fluid. Empty bladder and note exact time; discard specimen; exactly 60 and 120 min. later empty bladder again, measuring the volume and determining the urea (U) percentage in each. A specimen of blood for estimation of the urea percentage (B) is collected at any time during the test. U and B must be expressed in the same units.

Then if the output of urine (V) is greater than 2 ml./min. the clearance calculated from the formula UV/B is the maximum clearance and may be expressed as a percentage of the average normal maximum urea clearance, which is 75 ml./min.

If the output is less than 2 ml./min. the standard clearance is calculated from the formula $U\sqrt{V}/B$. This is not the actual clearance at which the patient's kidneys are working, but a theoretical figure calculated from the results. This is expressed as a percentage of the average normal standard urea clearance, which is 54 ml./min.

With children and adults of unusual size the clearances are multiplied by a factor to make them correspond to a surface area of 1·73 sq. metres, a maximum clearance by multiplying by $\frac{1\cdot 73}{A}$ and a standard clearance by $\frac{\sqrt{1\cdot 73}}{A}$, A being the surface area of the patient readily determined from the height and weight, by reference to published tables. The corrected clearances may then be compared with the average normal values for adults.

Over 70%	Normal function
70–40%	Mild deficit
40–20%	Moderate deficit
Below 20%	Severe deficit
Below 5%	Uræmic coma present or imminent

If proteinuria is severe or if there is generalized œdema:

1. Determine the serum protein and examine the electrophoretic pattern.
2. Determine the serum or plasma cholesterol.

3. If mechanism still obscure, determine protein content of œdema fluid.

The nephrotic syndrome is present when there is: (a) massive albuminuria; (b) massive œdema; (c) normal renal function; (d) normal blood pressure; (e) hypo-albuminæmia; (f) hypercholesterolæmia.

Ascorbic acid saturation test

8 a.m.	Patient empties bladder and is given 700 mg ascorbic acid by mouth.
12 noon.	Patient empties bladder. Specimen discarded.
2 p.m.	Patient empties bladder. Complete specimen collected and ascorbic acid content estimated within 1 hour.

The test is repeated daily until the patient excretes more than 30 mg of ascorbic acid in the 2-hour specimen.

Patients with ascorbic acid deficiency may take more than 7 days to reach saturation. The test is useful, since it also shows the response to therapy.

ROUTINE LABORATORY INVESTIGATION OF ADRENOCORTICAL DYSFUNCTION

If adrenocortical dysfunction is suspected, 17-hydroxycorticosteroids in 24-hour urine should be estimated; 17-oxosteroids are seldom required, but should be estimated if there is evidence of virilism. Subsequent investigations depend upon clinical evidence of: (a) primary or secondary adrenal hypo- or hyperfunction; (b) congenital adrenal hyperplasia or virilism. The following investigations mainly depend upon estimation of plasma cortisol; if facilities for these are not available the effects of stimulation or suppression may be assessed from the estimation of urinary 17-hydroxycorticosteroids.

Adrenal Hypofunction

In suspected acute adrenal hypofunction (usually manifest as Addison's disease) blood for estimations of electrolytes and cortisol* should be collected before treatment is begun (see p. 153). In chronic adrenal hypofunction, other tests are necessary.

Primary adrenal hypofunction

Screening Synacthen test

    9.00 a.m. 10 ml. blood* collected and 250 μg Synacthen in 1–2 ml. 0·9% NaCl given I.M.
    9.30 a.m. 10 ml. blood* collected.
    10.00 a.m. 10 ml. blood* collected.

Cortisol is estimated in all three blood specimens.

Normal response. 9.00 a.m. plasma cortisol exceeding 5 μg/100 ml.
    9.30 or 10.00 a.m. plasma cortisol should exceed 18 μg/100 ml. and show an increase of at least 7 μg/100 ml. above resting level.

The normal response excludes primary adrenal insufficiency and no other test is necessary.

Impaired response may be due to primary or secondary adrenal hypofunction and further tests are required.

Definitive Synacthen or ACTH test

9.00 a.m. 10 ml. blood collected and 500 μg Synacthen (or 50 I.U. purified ACTH) in 500 ml. 0·9% NaCl infused I.V. over 5 hr., during which blood is collected hourly.

Plasma cortisol is estimated on all 6 samples.

Normal response. 9.00 a.m. plasma cortisol should exceed 5 μg/100 ml.
    2.00 p.m. plasma cortisol should be within range 30–60 μg/100 ml.

---

* The plasma for cortisol estimations should be separated within 10 minutes of collection; if analysis cannot be performed at once the plasma MUST be stored in the frozen state.

Abnormal response. Primary adrenal insufficiency confirmed if plasma cortisol does not increase during test. No other test necessary.

Secondary adrenal insufficiency is probable if plasma cortisol rises to less than 30 µg/100 ml., but further tests are necessary.

Secondary adrenal hypofunction

Insulin stress test

The test is carried out after a 14-hour fast and glucose must be immediately available for intravenous injection should the severity of side-effects warrant it.

7.00 p.m. begin fast.
9.00 a.m. 12 ml. blood* collected . . . and 0·10 u soluble insulin/kg body wt. given I.V.
9.30 a.m. 12 ml. blood* collected every half hour till 11.00 a.m.

Plasma cortisol and blood glucose are estimated on each specimen.

Normal response. The resting plasma cortisol, the maximum increment above the resting plasma cortisol and the maximum plasma cortisol must exceed 5, 7 and 20 µg/100 ml. respectively.

Abnormal response. If response is subnormal in spite of adequate hypoglycæmia (i.e. blood glucose falling to below 40 mg/100 ml.) further tests are necessary.

Metopirone test

Day 1   24-hour urine (pre-metopirone)
Day 2   Give 750 mg. metopirone orally 6 hourly for 4 doses. 24-hour urine.
Day 3   24-hour urine (post-metopirone).

17-hydroxycorticosteroids measured in all 3 specimens.

Normal response. Urinary 17-hydroxycorticosteroids should increase by at least 10 mg/24 hr.

Subnormal response shows hypothalamo-pituitary deficiency. Confirm *if possible* with thyroid function tests, tests for other pituitary hormones (e.g. gonadotrophins, growth hormone, etc.) and tests for primary pituitary insufficiency by the lysine–vasopressin test.

Lysine–vasopressin (LVP) test

During this test vasoconstriction (perhaps cardiac ischæmia) and pallor may occur and on occasions intestinal hypermotility and the desire to defæcate. In these circumstances the infusion rate should be slowed till symptoms cease.

9.00 a.m. 10 ml. blood collected and 10 pressor units of LVP in 250 ml. 0·9% NaCl infused I.V. over 2 hours.

11.00 a.m. 10 ml. blood collected.

Blood specimens analysed for plasma cortisol.

Normal response. Maximum plasma cortisol increment should exceed 6·0 μg/100 ml.

Adrenal hyperfunction

Cushing's Syndrome may be established by the following preliminary test:

Diurnal rhythm and single-dose dexamethasone suppression tests

Day 1 9.00 a.m. 10 ml. blood collected.
11.00 p.m. 10 ml. blood collected and 2 mg. Dexamethasone given orally.
Day 2 9.00 a.m. 10 ml. blood collected.

Estimate plasma cortisol on all three blood specimens.

Normal response. Plasma cortisol, at 9 a.m. and 11 p.m. on day 1 and at 9 a.m. on day 2 should be not greater than 25, 5 and 3–6 μg/100 ml. respectively.

When a high plasma cortisol, an abnormal diurnal rhythm or a poor suppression is demonstrated the ætiology of the disease should be assessed from:

(*a*) Metopirone test (see p. 219).
(*b*) ACTH stimulation test (see p. 218).
(*c*) Multiple-dose dexamethasone test.

Multiple-dose dexamethasone suppression test

Day 1   24-hour urine collected.
Day 2 ⎱ Give 0·5 mg dexamethasone orally every 6 hours for
Day 3 ⎰   8 doses.
Day 4 ⎱ Give 2·0 mg dexamethasone every 6 hours for 8
Day 5 ⎰   doses.

17-hydroxycorticosteroids in 24-hour urine outputs are estimated each day of the test. For a *normal response* they decrease to less than 5 mg/day.

Primary adrenal carcinoma usually fails to respond; in hyperplasia there are usually responses to dexamethasone, ACTH and metopirone.

Adrenal hypersecretion due to extra-adrenal carcinoma is not suppressed with dexamethasone, but is often increased beyond resting levels with ACTH stimulation.

Adrenogenital syndrome

The investigation of the adrenogenital syndrome, whether due to a congenital enzyme deficiency or to an androgen-secreting tumour, initially requires estimation of urinary 17-oxosteroids. The enzyme deficiency may be established by the estimation in the urine of compound S and its metabolites or the 11-oxygenation index* (deficiency of 11-hydroxylase), pregnanetriol (deficiency of 21-hydroxylase) and pregnenolone and its metabolites (deficiency of 3β-hydroxy steroid dehydrogenase). Often more elaborate investigation may be necessary. Not infrequently, such enzyme deficiencies are associated with mineralocorticoid and electrolyte abnormalities.

The excretion of 17-oxosteroids is rapidly reduced by glucocorticoid therapy, but virilism due to a tumour is not usually responsive to dexamethasone suppression.

---

* The 11-oxygenation index is also abnormal when there is deficiency of 21-hydroxylase.

BLOOD ANALYSIS—

In using this table it is essential to realize that a small percentage of been extended to include this small number of people, the range would have and King (*Lancet*, 1953, i, 470) to meet this difficulty. They have compiled normal people respectively. The chances are very great that a finding out-finding falling outside the 80% range but inside the 98% range.

Determination.	Specimen required.	Normal values.
Urea	Blood	15–40 mg/100 ml. 15–50 children, old age 15–30 pregnancy
Total protein	Serum	6·3–7·9 g/100 ml.
Fibrinogen	Plasma	0·2–0·5 g/100 ml.
Bilirubin	Serum	0–0·75 mg/100 ml.
Glucose	Whole blood	Fasting { 80–110 mg/100 ml. by non-specific method 50–90 mg/100 ml. by specific method
Calcium	Serum	8·5–10·5 mg/100 ml.
Inorganic phosphorus	Serum or plasma separated within ½ hr of collection	2·0–4 mg/100 ml. 4–6 mg/100 ml. in infants
Uric acid	Serum	4–7 mg/100 ml. in males 2–5 mg/100 ml. in females
Cholesterol	Serum or plasma	100–220 mg/100 ml.

SOME NORMAL VALUES

normal people may have values outside the range given. Had the range been so wide as to be almost useless. Some attempt has been made by Wootton a table with two ranges covering these determinations in 80% and 98% of side the 98% range is abnormal, whereas the chances are much less for a

Increased in	Decreased in
Renal failure, (a) primary renal disease, (b) secondary to cardiac failure, (c) secondary to salt depletion, e.g., shock, diarrhœa, vomiting, intestinal obstruction, diabetic coma, fistulæ, Addison's disease, (d) fever, (e) anuria	Pregnancy Severe necrosis of liver
Water depletion, multiple myelomatosis (some cases), sarcoidosis (some cases), liver disease (some cases)	Protein deficiency, nephrotic syndrome, liver disease (some cases)
	Severe liver disease
Jaundice	
Diabetes, hyperthyroidism, Cushing's disease, acromegaly (some stages)	Functional hypoglycæmia, severe hepatic failure, Simmonds' disease, Addison's disease, tumour of islets of Langerhans
Hyperparathyroidism, some cases of multiple myelomatosis, occasionally in osteogenic sarcoma and sarcoid idiopathic hypercalcæmia of infancy, vitamin D over-dosage, immobilisation, excessive milk and alkali	Hypoparathyroidism, rickets (some cases), cœliac disease and renal rickets, osteomalacia associated with calcium or vitamin D deficiency, or the sprue syndrome, hypoproteinæmia
Renal failure, hypoparathyroidism	Rickets (some cases), vitamin D over-dosage, hyperparathyroidism
Uræmia, leukæmia, gout	
Nephrotic syndrome, myxœdema, cretinism, pregnancy, nephritis (some cases), diabetes (some cases), arteriosclerosis (some cases), certain forms of xanthomatosis	The statistically demonstrable fall in hyperthyroidism is useless for diagnosis in the individual patient

BLOOD ANALYSIS—

Determination.	Specimen required.	Normal values.
Sodium	Serum	130–145 mEq/l.
Potassium	Serum separated within 2 hrs of collection	3·8–5·4 mEq/l.
Chloride	True plasma	96–108 mEq/l.
Bicarbonate	True plasma	24–34 mEq/l.
Amylase	Plasma or serum	80–180 units/100 ml. by Somogyi method
Alkaline phosphatase	Serum or plasma separated within 2 hr of collection	3–13 units/100 ml. 15–20 units/100 ml. in infants
Acid phosphatase	,, ,,	0–4 units/100 ml.
Transaminase G.O.T.	Serum	5–30 Reitman–Frankel units/ml.
G.P.T.		0–25 Reitman–Frankel units/ml.
Lactic dehydrogenase		96–290 i.u./l.
Aldolase		0·9–2·5 u/l.
Creatinine phosphokinase		less than 1 u/l.
Cortisol	Plasma	9 a.m. 6–25 μg/ml. 11 p.m. less than 5 μg/ml.
Protein-bound Iodine	Plasma	3–8 μg/100 ml.

SOME NORMAL VALUES—continued

Increased in	Decreased in
Water depletion, certain forms of nephritis, over-treatment of Addison's disease with DOCA or other mineralocorticoids	Salt depletion, e.g., shock, diarrhœa, vomiting, intestinal obstruction, diabetic coma, fistulæ, Addison's disease
Severe salt depletion, e.g., shock, diarrhœa, vomiting, intestinal obstruction, diabetic coma, fistulæ, Addison's disease, renal failure (some cases)	Treatment of salt depletion with saline, during treatment of diabetic coma, Cushing's syndrome, cortisone and ACTH therapy, familial periodic paralysis
Water depletion, certain forms of nephritis, after ureteric transplantation, overtreatment of Addison's disease with DOCA, renal acidosis of infancy, biliary fistulæ, pancreatic fistulæ	Salt depletion, e.g., shock, diarrhœa, vomiting, intestinal obstruction, diabetic coma, fistulæ, Addison's disease, certain forms of nephritis
After ingestion of alkalis, vomiting acid gastric juice, CO_2 intoxication	Non-respiratory bicarbonate deficit, especially renal failure, diabetic coma, after ureteric transplantation, renal acidosis of infancy, CO_2 deficit
Acute pancreatitis, some cases of perforated peptic ulcer, salivary duct calculus, some cases of mumps	
All forms of jaundice except hæmolytic (more general in obstructive jaundice), all bone diseases associated with osteoblastic activity, e.g., osteomalacia, rickets, etc., Paget's disease, hyperparathyroidism	
Metastasizing prostatic carcinoma	
Especially cardiac infarction, also liver damage	
Especially liver damage, also cardiac infarction	
Cardiac infarction, liver damage	
Liver damage	
Cardiac infarction, primary disease of muscle and secondary to damage to nerve tissue	
Cushing's Disease, hyperthyroidism, pregnancy	Addison's Disease, Panhypopituitarism, Primary and secondary hypothyroidism.

INDEX

ABORTION, habitual, 157
Acatalasia, 92
Acetanilide, 168
Acetest, 132, 211
Aceto-acetic acid, 131, 132, 184, 210
Acetylcholine, 79
Acetylcoenzyme A, 127, 131
Achlorhydria, 112–114, 182
Acidæmia, 19, 138
Acid–base balance, 19 et seq.
  assessment of, 28
  changes in, 26, 27
  classification of, 23
  disturbances of, 23 et seq.
  effect of anæsthetics, 24
  effect of barbiturates, 24
  in pyloric stenosis, 27
  ions in, 19
Acidity, gastric, 2, 110
Acidosis, 19
  hyperchloræmic, 18, 25, 26, 164
  renal tubular, 8
Acid secretory rate, by gastric juice, 115
Acromegaly, 100, 130, 135, 155
ACTH, 45, 147, 152, 154, 155, 220
  test, 218
Addison's disease, 14, 52, 59, 63, 129, 158, 217
  chronic, 153
  crisis in, 153
  electrolyte patterns in, 63, 153
Adenosine triphosphate (ATP), 195
Adenylate kinase, 163
Adrenalectomy, 52
Adrenal hydrogenation:
  11-hydroxylase, 150, 151
  17-hydroxylase, 150, 151
  21-hydroxylase, 150, 151

Adrenal, cortical hormones, 129, 146 et seq.
  carcinoma, 143, 221
  disease, 148 et seq.
  function, investigation of, 217–220
  hyperplasia, congenital, 150
  insufficiency, 153, 217–220
  medulla, 154
  metabolism, 147
  secretion rate, 150
  steroids, 146, 151
  tumour, 152, 153, 154, 183, 221
Adrenalin, 127, 128, 154
Adrenogenital syndrome, 150, 221
Afibrinogenæmia, 92
Albers-Schonberg disease, 101
Albinism, 164, 165
Albumin, ^{131}I-labelled, 84
  human, treatment with, 44
  Osmotic pressure of, 37
Albustix, 207
Alcohol, poisoning by, 185
Alcoholism, chronic, 181
Aldolase, 71, 177
Aldosterone, 7, 36, 41, 44, 77, 146
  metabolites, 44, 147
  role of œdema, 41, 44
Aldosterone antagonists, 47
Aldosterone secreting tumour, 41, 154, 183
Aldosteronism, 17, 25, 46, 154
Alkalæmia, 19
Alkali reserve, 1, 22
Alkalosis, 19, 24, 97
Alkaptonuria, 163–165
Alleles, 161
Amino acidurias, 18, 164
ε-Aminocaproic acid, 200
β-Aminoisobutyric acid, 166

INDEX

δ-Aminolævulic acid (ALA), 169
  synthetase, 191, 192
Amino-transferases, 173–175
Ammonia, 80
  coefficient, 27
Amylase, 172
  in urine, 171
  plasma, 116
Amyloid disease, 16, 40
Amyloidosis, 91
Amylo-1-6-glucosidase, 167
Amylo-(1:4–1:6)-transglucosidase, 168
Anacidity, 115
Anaemia, aplastic, 198
  hæmolytic, 195–198
  hæmolytic, auto-immune, 197
  hæmolytic congenital nonspherocytic, 196
  hypochromic, 113, 115
  megaloblastic, 113, 188–191
  pernicious, 113, 115, 116, 182, 187, 195
  sickle-cell, 198
Anæsthetics, effect on acid–base balance, 24
Analbuminæmia, 92
Androgens, 146, 148, 150, 155, 156
Androstenedione, 146, 155
Antibodies, cold auto-, 197
  Donath–Landsteiner, 197
Antidiabetic drugs, 140
Anti-diuretic hormone, 7, 40, 44, 77
Antipyrine, 168
Antithyroid drugs, 142
Anuria, 15
Apoferritin, 188
Arginine, 18, 165
Arginino-succinic acid, 166
Arrhenoblastoma, 158
Aschheim–Zondek test, 158
Ascites, 39, 42
  treatment of, 44–47
Ascorbic acid, 201
  saturation test, 217
Ashby agglutination test, 194
Atheroma, 93, 203
Azobilirubin, 67
Azotæmia, 13, 14
  extra-renal, 14

Azotæmia, in intestinal hæmorrhage, 15
  post-renal, 14, 15
  pre-renal, 14, 53
  renal, 14

Barbiturates, effect on acid–base balance, 24
  estimation of, 186
Basal metabolic rate (*see* B.M.R.)
Base reserve, 22
Bence-Jones protein, 86, 90, 91, 92
Benedict's test, 208, 209
Betamethasone, suppression by, 149
Bicarbonate, 1
  concentration, 26
  deficit, 25
  excess, 25, 54
  excretion of, 8
  plasma, 22, 28, 46
  reabsorption of, 7, 8, 18
  "standard", 28
Bile pigments, 64, 67–69, 72, 198, 211, 212
Biliary, fistula, 55, 60
  deficiency of, 119
Bilirubin, 64, 211
  conjugated, 64, 66
  determination of, 67
  formation and fate of, 64, 65
  plasma, 65
  tests for, 69
Bisalbuminæmia, 92
Blood, collection of, 206
  hydrostatic pressure of, 36
Blood glucose, 126
  hormonal control, 128
Blood groups, 160, 161
Blood lipids, 135
Blood proteins, 161, 162
Blood sugar, 126, 136
B.M.R., 143
  in leukæmia, 144
Body fluids, distribution of, 33, 34, 35
  ionic composition of, 2
Body water, determination of, 33
Bone, fibrocystic diseases of, 105
  mechanism of formation, 94
  tumours in, 101, 102

INDEX

Bone crystals, 94
Bone disease, 96, 170, 173
  classification of, 102
Borst diet, 16
Bradshaw's test, 208
Bromsulphthalein test, 70
Buffer, base, 20, 29
  capacity, 20
  carbonic acid–bicarbonate system, 21
  hæmoglobin–hæmoglobinate system, 22
  mechanism, 19
  systems of the body, 20
Burns, 40

Cæruloplasmin, 182
Calciferol (vitamin D), 93, 96, 98, 99, 100, 103, 104
  deficiency, 99, 100
  resistance, 99, 100
Calcification, in kidney, 97
  mechanisms of, 93
  pathological, 93
Calcitonin, 97
Calcium, 202
  balance, 108
  serum, 96
  urinary, 97, 98, 99, 105, 106, 108
Calculus, 93
  biliary, 73
Capillary permeability, 36
Carbohydrate starvation, 135
Carbonic anhydrase, 8, 46
Carbon monoxide, 185
Carbon tetrachloride poisoning, 170
Carcinoma, gastric, 112, 113
  of the breast, 101
  of the pancreas, 117
Carcinoid tumours, 125
Carcinomatosis, 98, 176
Cardiac failure, 14
  congestive treatment of, 47
  glomerular filtration in, 41
  in Paget's disease, 107
  œdema of, 40
  secondary hyperaldosteronism in, 41
Cerebrospinal fluid, 33
  bile pigment in kernicterus, 184

Cerebrospinal fluid, glucose in, 180
  pigment in, 180
  protein in, 180
  routine investigation of, 179
Childhood, adipose gynism in, 204
  electrolyte deficiencies in, 62
  gynandrism in, 204
  hypercalcæmia in, 98
Chlorides, 181, 212
  estimation of, 57
  shift, 22
  urinary, 57
Cholesterol, and thyroid disease, 146
  in jaundice, 71
  plasma, 17, 71, 135, 146, 205
Cholinesterase, 71, 162, 197
  deficiency of, 162, 177, 183
Chorioncarcinoma, 157, 158
Chorionic gonadotrophin (CG), 158
Chromium sesquioxide, 108
Circulatory failure, 40
Cirrhosis, of the liver, 42, 69, 76, 77, 89
  post-hepatitis, 76
Clearance test, 8, 14, 145
Clinistix, 209
Clinitest, 209
CO_2, excess of, 24
  deficit, 24
Cobalamin, clearance of, 10
Cœliac disease, 119, 121, 122
Colitis, ulcerative, 85, 118
Collagen disease, 89
Coma, 24
  chemical pathology of, 184
  diabetic, 25, 135, 140, 179, 184
  gastric lavage in, 185
  hepatic, 79, 80
  of hypoglycæmia, 129, 185
Compound S, 149, 150
Conjugate bases, 20
Conn's syndrome, 41, 154, 183
Copper, in hæmopoiesis, 188
  metabolism of, 182
Coproporphyrin III, 77
Coronary, infarction, 173–176
  disease, 203
Corticosterone, 146, 147
Cortisol, 146, 147

Cortisol, in plasma, 148, 217–220
  secretion rates of, 150
Counter-current system, 5–7
Creatine phosphokinase, 177
Creatinine, blood, 14
  clearance, 10
Cretinism, 146, 164, 165, 179, 205
Crushing injuries, 15
Cryoglobulinæmia, 89
Cushing's syndrome, 25, 102, 129, 130, 134, 152, 220
Cyanocobalamin, 189
Cystathionine, 165
Cystine, 18, 164, 165
Cystinosis, 164
Cystinuria, 164

Davenport diagram, 30
Dehydration, in diabetic coma, 138
Dehydro*epi*androsterone, 146, 155
Dehydrogenases, 71, 176
  glucose-6-phosphate, 71, 176
11-Deoxycortisol, 149, 150
Deuteroporphyrin, 123
Dexamethasone, suppression by, 149, 220, 221
Dextran, 44, 85
Diabetes, 16, 25, 126–141, 203
  causes of coma, 135
  detection of, 132, 140
  insipidus, 18
  lypodystrophy in, 204
  mellitus, causes of, 129, 134
  metabolism in, 131
Diabetic coma, 135 *et seq.*, 179, 184, 185
  electrolyte patterns of, 63, 138 *et seq.*
Diarrhœa, 14
  electrolyte patterns of, 63
  fatty, 118 *et seq.*
Di-iodotyrosine, 142
Diodrast clearance, 9
Disseminated lupus erythematosus, 16, 89
Diuretics, 8, 25, 45–47, 61, 99
  mercurial, 45
  salt deficiency during therapy with, 52
  thiadiazine, 46, 99

Dropsy, 38
Dyspepsia, 113

Echinococcus in tuberculous infections, 93
Ehrlich's aldehyde reagent, 169, 211, 213
Ehrlich's test for porphobilinogen, 213
Ehrlich's test for urobilinogen, 211
Electrolytes, 1
  absorption by renal tubules, 5–8, 27, 45–47
  analyses of, 57
  deficiencies in infancy and childhood, 62
  patterns in Addison's disease, 63, 153
  patterns in chronic nephritis, 63
  patterns in diabetic coma, 63, 138 *et seq.*
  patterns in diarrhœa, 63
  patterns in ketosis, 63
  patterns in toxic vomiting of pregnancy, 63
  treatment of deficiency, 56
Electrophoresis, 81, 162 *et seq.*
  Immuno-, 83
  of FIGLU, 191
Emphysema, 24
Enteritis, regional, 85
  gastro-, 62
Enteropathy, protein-losing, 84, 85
Enzymes, 162
  activity in body, 178
  activity in impaired renal function, 178
  activity, international units of, 172
  assay of, 172
  glycolytic, 176
  in normal blood, 170
  iso-, 171
Enzymology, clinical, 170
  clinical limitations of, 177
Erythropoiesis, 187
Erythropoietin, 188
Exchange, transfusion, 78, 184
  resin for gastric function, 114
  resins, 45
Exophthalmos, 143

INDEX

Extracellular fluid, 33 et seq., 50–55, 138, 153
   calculation of, 34
   determination of, 33
   loss of, 51 et seq.
Exudates, 39

Familial periodic paralysis, 183
Fanconi syndrome, 18, 100, 105
Fat, absorption, 117 et seq.
   absorption defect, laboratory examinations in, 122
   deposition, abnormalities of, 204
   fæcal, 118
   transport of, 202
Fatty acids, non-esterified, 131, 135, 140
Favism, 167
Ferritin, 188
Fibrin, 198–200
Fibrinogen, 198–200
Fibrinolysis, 198–200
Fibrocystic disease, congenital, 119
   sweat test for, 119
FIGLU, 191
   electrophoresis of, 191
Fistulæ, intestinal, 118
Fluid balance, 49
Folic acid, 188, 189–191
Folinic acid, 190
Follicle-stimulating hormone, 157, 158
Formimino glutamic acid, 189–190
Fragilitas ossium, 101
Friedman test, 158
Fructosuria, 166

Galactokinase, 167
Galactosæmia, 18, 166
   congenital, 166, 179, 208
Galactose, 166
   -1-phosphate, 166
   -1-phosphate uridyl transferase, 166, 167
   tolerance tests, 71
   transferase, 166, 167
Gargoylism, 183
Garrod, 160, 168
Gastric, analysis, 110

Gastric, carcinoma, 85, 112
   function, 110 et seq., 190
   insulin test, 114
   lavage, 51, 59, 185
   residuum, 112
   ulcer, 115
Gastric juice, 2, 3, 172
   acid in, 2
   acid secretory rate, 115
   loss of, 2, 54
Gastritis, hypertrophic, 85
Gastro-enteritis, 62
Gastro-intestinal hæmorrhage, 123
Gastro-intestinal secretions, 49, 50
   loss of, 50, 52, 54
Gaucher's disease, 183
Genes, 160
Genetics, biochemical, 160
Gerhardt's test, 132, 210
Gigantism, 130, 155
Gilbert's disease, 67
Globin, 191, 193
Globulin, 81, 82
γ-Globulins, 72, 76, 77, 79, 82, 85 et seq.
   in infants, 78
Glomerular filtration, in cardiac failure, 41
   rate, 5, 9, 10, 15, 17
   rate in the nephrotic syndrome, 17
Glucocorticoids, 146, 147
Gluconeogenesis, 126
Glucose, concentration in blood, 136
   in cerebrospinal fluid, 180
   in urine, 137, 208–210
   -6-phosphatase, 127, 167, 173
   -6-phosphate, 127
   -6-phosphate dehydrogenase deficiency, 167, 195, 196
   Tm, 11, 137
   tolerance test, 71, 133
Glucuronyl transferase, in infancy, 78
Glutamic acid, 80
Glutamine, 80
Glutathione, 196
   dehydrogenase, 176
   reductase, 176
Gluten, 122

INDEX

Glycogen, storage disease, 167, 168, 173
Glycolytic pathway, 195
Glycosuria, 17, 132
   renal, 105, 134, 164, 166
Gout, 47, 204
Growth hormone, 100, 155
   role in diabetes, 130
Gut, obstruction of, 51
Gynandrism, in childhood, 204
Gynism, adipose, in childhood, 204

Hæmatemesis, 77, 123
Hæmatocrit, 55, 56
Hæmochromatosis, 169, 189
Hæmodialysis, 16
Hæmoglobin, 161, 162, 191 *et seq.*
   biosynthesis of, 191
   fœtal, 162, 193
   metabolism of, 194
   sickle-cell, 162, 194
   synthesis and degradation of, 191–193
Hæmoglobin A, 162, 193
Hæmoglobinuria, 12
   paroxysmal nocturnal, 197
Hæmolytic disease, 176, 196 *et seq.*
   of infants, 77, 161, 175
Hæmorrhage, 69, 77
   gastro-intestinal, 15, 123
Hæmosiderosis, 194
Haptoglobins, 84
Harrison's test, 211
Hartnup disease, 18, 166
Heinz bodies, 196
Hemastix, 125, 214
Hematest, 124, 125, 214
Heparin, 204
Hepatitis, acute, 74, 174, 175
   chronic, 76
Hepatolenticular degeneration, 18, 182
Hereditary spherocytosis, 197
Hirsutism, 156
Histamine test, 113
   augmented, 113
Hodgkins' disease, 176
Hogben test, 158
Homogentisic acid, 132, 163

Hormones, adrenal cortical, 146 *et seq.*
   anti-diuretic, 7, 40, 44, 77
   follicle-stimulating, 157, 158
   growth, 100
   luteinizing, 156–158
   mineralocortical, 7
   parathyroid, 98, 105
Hydatidiform mole, 158
Hydrocephalus, 179
Hydroxy apatite, 94
Hydroxycobalamin, 189
17-Hydroxycoticosteroids, 148, 153, 217
11β-Hydroxylase, 150, 151, 221
   inhibitor, 149
17-Hydroxylase, 150, 151, 221
21-Hydroxylase, 150, 151, 221
5-Hydroxytryptamine, 125
Hyperaldosteronism, 41, 154, 183
Hyperbilirubinæmia, congenital, 67, 169
Hypercalcæmia, 97, 98
   idiopathic, 98
   in childhood, 98
Hypercalciuria, 98, 99
   idiopathic, 99
   in Paget's disease, 99
Hyperchlorhydria, 111
Hypercholesterolæmia, 169
Hyperglobulinæmia, 89
Hyperglycæmia, 129
Hyperparathyroidism, 98, 100, 101, 105, 108
   diagnosis of, 105–107
Hyperpituitarism, 155
Hypertension, 13
Hyperthyroidism, 129, 134, 142, 145
Hyperuricæmia, 169
Hypocalcæmia, 99
   in chronic sepsis, 99
   in liver disease, 99
   in nephrotic syndrome, 99
   in protein deficiency, 99
Hypochlorhydria, 111
Hypogammaglobulinæmia, 86 *et seq.*
   classification of, 88
Hypoglycæmia, 71, 79, 128, 129
   of coma, 129, 185

INDEX

Hypoparathyroidism, 97, 99
Hypophosphatasia, 169
Hypoproteinæmia, 16, 17, 40, 84
   idiopathic, 84, 85
Hypoprothrombinæmia, 72, 76, 112
Hypothesis, one gene-one enzyme, 160

Ictotest, 211
Ileitis, regional, 103
Immunochemical methods, 12, 83, 84, 129, 155, 157, 158
Immunoglobulins, 85
   classification of, 86
Infancy, electrolyte deficiencies in, 62
   glucuronyl transferase in, 78
   haemolytic disease in, 77, 161, 175
   jaundice in, 78
Infection, *C. Welchii*, 15
Insulin, 128 *et seq.*, 149
   effect of, on cortisol secretion, 149, 219
   inhibitors of, 129, 130
   in plasma, 129
   gastric function test, 114
   phosphate after, 100
International units, of enzyme activity, 172
Interstitial fluid, 20, 33, 36–38
   exchange of fluid with plasma, 36–38
   from the liver, 42
Interstitial space, 43
Intestinal obstruction, 14, 51, 59
   fistulæ, 118
   function, 51
   loss of secretions, 52, 54
Intracellular fluid, 3, 33, 35, 51 *et seq.*
   cations in, 3, 34
   determination of volume, 33
   loss of, 53
Intrinsic factor, 116, 189, 190
Inulin, clearance, 9
Iodine, protein-bound, 143, 144, 155
Ion exchange resins, 25, 45
Iron, binding capacity, 189
   in plasma, 188
   metabolism of, 188
Islet-cell tumours, 116, 129

Isochlorhydria, 111
Isocitrate dehydrogenase, 176
Iso-enzymes, 171
Isoleucine, 166
Isosthenuria, 13

Jaundice, 64, 118, 119
   cholestatic, 76, 79
   cholesterol in, 71
   classification of, 65
   congenital non-hæmolytic pre-hepatic, 66–70, 73, 78, 79, 198
   differential diagnosis of, 72
   Dubin-Johnson, 67
   due to drugs, 66, 79
   enzymes in, 71
   hæmolytic pre-hepatic, 65, 66, 67, 73, 173–176
   hepatic, 65, 66
   in infancy, 78, 175
   non-hæmolytic pre-hepatic, 67, 74, 169
   obstructive, 65, 173, 175
   phosphatase in, 70
   pigment excretion in, 68, 69
   post-hepatic, 65
   pre-hepatic, 65

Kala-azar, 89
Kernicterus, 78, 184
   bile pigment in, 184
17-Ketogenic steroids (*see* 17-Oxogenic steroids)
Ketonæmia, 137, 138
Ketone bodies, 131, 135, 137
   in blood, 137
   in urine, 137, 210–211
Ketonuria, 132
Ketosis, 25, 131, 132, 133, 135, 137
   electrolyte patterns in, 63
17-Ketosteroids (*see* 17-Oxosteroids)
Kidney, calcification in, 97, 105
   physiology of, 5
   polycystic, 13
   tuberculous, 13
   tumour of, 13
Kolff apparatus, 16
Krebs cycle, 80, 127, 131
Kunkel zinc sulphate test, 72, 87
Kwashiorkor, 85

Lactic dehydrogenase, 71, 171, 175, 176
Lactose, 132
Lævulose, 132
  tolerance test, 71
"Lag storage curve", 133
Lange colloidal gold curve, 180
LATS, 143, 145, 154
Leucine, 166
  in urine, 71, 76
Leukæmia, 176
  B.M.R. in, 144
Lipase, 172
Lipidoses, 182, 183
Lipids, in blood, 17
Lipodystrophies, 204
Lipoproteins, 202, 203
Liver, acute necrosis of, 76
  cirrhosis of, 42, 69, 76, 77, 89
  damage (*see* hepatic jaundice)
  damage, neuropathy in, 79, 80
  disease, 173–175
  failure, 70, 71, 76, 77, 79
  interstitial fluid from, 42
  neurological disturbances of chronic disease, 185
Liver function tests, 69–73
  classification of, 69
Lœser's zones, 104
Luteinizing hormone (LH), 156–158
Lymph, 37
Lymphatic drainage, 37
Lymphogranulomatosis, intestinal, 85
Lysine, 18
Lysine-vasopressin test, 149, 220

Macroglobulin, in urine, 12
Macroglobulinæmia, 89
Magnesium, 97
  deficiency of, 62, 97
Malabsorption syndrome, 119 *et seq.*
Mannitol, clearance, 9
"Maple syrup urine disease", 166
Melæna, 123
Meningitis, 180, 181
Menopause, 102, 158
Menstruation, 156, 158
Mercaptans, 80

Mesoporphyrin, 123
Metabolism, inborn errors of, 160
Metopirone, 149, 219, 220
Methæmoglobinæmia, 168
Methionine sulphoxide, 80
Milliequivalents, 1
Millimoles, 1, 4
Milli-osmols, 1, 4
Mineralocorticoids, 41, 147, 150, 154
Monophosphate shunt, 195
Morphine, effect on acid-base balance, 24
Mucopolysaccharides, 161
Mumps, plasma amylases in, 116
Muscle, diseases of, 176, 177
Myeloma globulin, 90
Myelomatosis, multiple, 89, 98
Myocardial infarction, 71, 92, 173–176
Myxœdema, 144, 146, 203, 205

NADH, 195
NADPH, 131, 195
Nephritis, 13, 16, 39, 40, 41, 42
Nephrocalcinosis, 18, 99, 105
Nephron, 5
Nephrotic syndrome, 12, 16, 40, 43, 205, 217
  hypocalcæmia in, 99
  glomerular filtration rate in, 17
  treatment of, 44 *et seq.*
Neuropathy, in liver damage, 79
Nicotinamide, 202
Niemann–Pick disease, 182, 183
Nitrite, 168
Nitrogen, non-protein, 14
Nor-adrenalin, 154, 165
Normoblasts, basophil, 187
  orthochromatic, 187
  polychromatic, 187
5'-nucleotidase, 173
Nutritional deficiencies, 201

Obesity, 204
Obstruction, of large intestine, 52
  of small intestine, 51
Occult blood, 123–125
  isotope method of determination of, 124
  spectroscopic tests, of, 123

INDEX

Occultest, 124, 213
Occular fluids, 33
Œdema, 36 et seq., 84
  angio-neurotic, 39
  classification of, 39
  congestive, 40
  hypoproteinæmic, 40
  inflammatory, 39
  lymphatic, 41
  of cardiac failure, 40
  role of aldosterone, 41, 44
  sodium retention of, 41
  treatment of, 44–48
Œsophageal obstruction, 50
  varices, 43, 76, 77
Œstriol, estimations, 157
Œstrogens, 155, 156
  effect on phosphatase, 102
Oliguria, 15
  salt and water balance, 60
One-carbon-atom fragments, 190
One gene-one enzyme hypothesis, 160
Open heart surgery, 31
Organophosphorus insecticides, 177
Ornithine, 18, 165
Orthostatic proteinuria, 12
Ortho-tolidine test, 124, 214
Osazone, 133
Osmoceptors, 41
Osmolality, 4
Osmolar clearance, 11
Osmolarity, 4
Osteoblastic activity, 94, 101
Osteoclasts, 94, 95
  activity of, 107
Osteoid, 94, 102
Osteomalacia, 18, 103, 104
Osteoporosis, 102, 103
Oxaluria, 169
17-Oxosteroids, 148, 153, 217
17-Oxogenic steroids, 148, 153
11-Oxygenation index, 150, 221

Paget's disease, 107
  cardiac failure in, 107
  hypercalciuria in, 99
PAH clearance, 9
PAH T_m, 11
Pancreas, carcinoma of, 117
Pancreas, islet cell tumours of, 116
Pancreatectomy, 130
Pancreatic deficiency, 119
  disease, 116 et seq.
  juice, 3, 172
Pancreatitis, acute, 116, 177
Panhypopituitarism, 129, 130, 146, 154, 158
Para-aminohippuric acid (see PAH)
Paralytic ileus, 59
Paramyloidosis, 90, 91
Para-phenylenediamine oxidase, 182
Paraproteinæmia, 89
Parathormone, 97
P_{CO_2}, 22, 138
  determination of, 28
Pentose, 132, 133
Pentosuria, 160, 166
Pepsinogen, 116, 172
Pericardial, effusions, 39
  fluids, 33
Peripheral circulatory failure, 14, 53, 140
Peritoneal dialysis, 16
  fluids, 33
pH, intracellular, 23
  blood, determination of, 28
Phæochromocytoma, 154
Phenacetin, 168
Phenistix, 213
Phenyl acetic acid, 163, 165
Phenylalanine, 163–165
  3:4 dihydroxy (DOPA), 165
Phenylhydrazine, 196
Phenylketonuria, 163–165, 179, 213
Phenyllactic acid, 163, 165
Phenyl pyruvic acid, 163, 165
  2:5 dihydroxy-, 165
Phosphatase, 70, 72, 75, 76, 77, 79, 93, 94, 100–102, 107, 171, 173
  effect of œstrogens, 102
  glucose-6-, 127, 167, 173
  placental alkaline, 163
  plasma acid, 101, 163
  plasma alkaline, 70, 100
Phosphate, 14, 97
  after insulin, 99
  clearance, 107
  in acromegaly, 100

INDEX

Phosphate, in urine, 27
   plasma inorganic, 14, 99, 100, 105, 140, 164, 202
   reabsorption in renal tubules, 97
   role in body, 99
Phosphoglucomutase, 127, 163, 167, 176
Phosphohexoisomerase, 176
Phosphorylase, 94, 127, 168
p-hydroxyphenyl pyruvic acid, 165
Pituitary disease, 148 *et seq.*, 154
Placental, alkaline phosphatase, 163
Plasma, cations in, 35
   determination of volume, 34
   exchange of fluid with interstitial fluid, 36
   osmotic pressure of, 37
   pyruvate, 181
   true, 207
   volume of, 33, 34
Plasmapheresis, 37
Plasma proteins, 17, 37, 56, 81
   sedimentation constants, 88
Plasmin, 198, 199
Pleural, effusions, 39
Poisoning, alcoholic, 185
   aspirin, 24
   carbon tetrachloride, 170
   investigation of, 185
   mercury, 17
   salicylate, 24
Poisons, elimination of, 16
Polyarteritis nodosa, 16
Polyostotic fibrous dysplasia, 105
Polyvinyl pyrrolidone, 85
Porphobilinogen, 169, 213
Porphyria, 160, 168
   acute intermittent, 168, 179, 213
   congenital, 160, 168
Portacaval anastomosis, 43, 69
Potassium, 17, 60
   concentration in cells, 35
   deficiency of, 17, 47, 60, 61, 164
   deficiency, clinical aspects of, 61
   deficit due to mercurial diuretics, 46, 47
   estimation of, 61
   excretion of, 8
   intravenous administration of, 61
   plasma, in diabetes, 139

Pregnancy, 157
   electrolyte patterns in vomiting of, 63
   vomiting of, 25
Pregnanediol, 157
Pregnanetriol, 150, 169, 221
Pregnanetriolone, 169
Pregnenolone, 150, 221
Primaquine, 167, 196
   sensitivity, 176
Progesterone, 157, 169
   hydroxylations of, 150, 151
Prostate, carcinoma of, 101, 102, 171
Protein, C-reactive, 92
   in C.S.F., 180
   in urine, 11, 207
   tissue fluid, 37
Proteinuria, 11, 12, 17
   mechanisms of, 17
   orthostatic, 12
Prothrombin, 71, 198
Protoporphyrin, 123, 191
Protoporphyrin IX, 191, 192
Pseudo-fractures, 104
Pteroylglutamic acid, 189, 196
Pyelonephritis, chronic, 17
Pyloric obstruction, electrolyte patterns in, 63
Pyloric spasm, 110, 112
Pyloric stenosis, 51, 54, 59
   acid–base balance in, 27
Pyruvate, plasma, 181

Radio-iodine tests, 145, 155
Red cells, 187, 190, 195, *et seq.*
   breakdown of, 195, 196, 197, 198
   integrity of, 195
   life span of, 198
Reducing substances, 132, 208, 209
Renal, clearances, 8
   control of pH, 23
   disease, 12–15, 25
   failure, 13, 25
   function, enzyme activity in, 178
   function tests, 12–15
   glycosuria, 105, 133, 134, 164, 166
   insufficiency, 10, 12
   medulla, 11
   papilla, 11, 15
   plasma flow, 10, 17

INDEX

Renal, stone, 15, 99, 106
  threshold, for glucose, 133
  tubular acidosis, 8
  tubular defects, 17, 18
  tubular function, 5–8
Renin, 154
Respiratory, control of pH, 23
  passages, obstruction of, 24
  quotient, 143
Reticulocyte, 187
Reticuloses, 89
Rheumatoid arthritis, 197
Rickets, 103–105
  renal, 104
  vitamin D resistant, 18, 164
Rothera's test, 132, 210

Salicylate, therapy, 132
Salicyl-sulphonic acid test, 207
Saline, therapy, 55 et seq.
  calculation of, 55
Salt, deficiency, 14, 51 et seq.
  deficiency during diuretic therapy, 52
  and water deficiency, 58
  treatment of, 56
Sarcoid, 89
Sarcoma, osteogenic, 93, 98
Schilling test, 190
Schlesinger's test, 169, 212, 213
Scleroderma, 118
Secondary hyperaldosteronism, 41, 44
Shock, 15
Sickle-cell, anæmia, 161, 193, 194, 198
  hæmoglobin, 161, 162
  "trait", 161
Siderophilin (see transferrin)
Simmond's disease (see Panhypopituitarism)
Small intestine, function of, 117
  obstruction of, 51
Sodium, deficiency of, 47, 56, 138, 153
  excretion of, 7
  reabsorption of, 7
Spectroscopic tests, 123
  for occult blood, 123
Sphingomyelin, 183

Spinal cord, sub-acute combined degeneration of, 182, 190
Spironolactone, 47
Sprue syndrome, 85, 103
  non-tropical, 119 et seq.
Starvation, 25
  carbohydrate, 135
Steatorrhœa, 118 et seq.
Stein–Leventhal syndrome, 156
Stem cell, 187
Stercobilin, 64
Steroid hormone, biosynthesis and metabolism of, 151, 156
  conjugation of, 147, 148
Stomach, acute dilatation of, 51, 59
Succinyl coenzyme A, 191, 192
Sulphonamide, 15
Sulphydryl enzymes, 45
Suxamethonium, 161, 162
  sensitivity, 162
Sweat test, for fibrocystic disease, 119
Synacthen Test, 218
Synovial fluids, 33
Systemic lupus erythematosus, 89

Test meal, gastric, 110
  histamine, 113
Testicular homones, 155
Testosterone, 146, 156
Tetany, 17, 62, 97, 179
Tetrahydrofolic acid, 189, 190
Thalassæmia, 193, 194, 198
Thiamine, 181, 202
Thiocyanate, 142
Thoracic injuries, 24
Thrombin, 198–200
Thromboplastin, 198–200
Thrombosis, 198
  of renal vein, 17
Thymol turbidity test, 72
Thyrocalcitonin, 98
Thyroid, physiology of, 142
  hormones, protein binding of, 143
Thyrotoxicosis, 102, 144, 146
Thyrotrophic hormone, 143, 145, 146
Thyroxin, 142
Tissue tension, 37
Titratable acidity, 27, 110

INDEX

Tolbutamide, 140
Transaminases, 71, 75, 76, 171, 173, 175
Transferrin, 12, 83
Transudates, 39
Trauma, 15
Trichinella spiralis, 93
Tri-iodothyronine, 142, 165
Trypsin, 172
  in stools, 119
Tubeless dye test, 114
Tubular excretory capacity, maximum, 10
Tubular reabsorptive capacity, maximum, 10
Tumour, aldosterone-secreting, 41, 154, 183
  carcinoid, 125
  granulosa-cell, 158
  islet cell, 116
  of adrenal medulla, 154
  testicular, 158
  in bone, 101, 102
Tyrosine, 165
  in urine, 71, 76

UDPG-glycogen transglucosylase, 168
Ulcer, duodenal, 112, 115
  gastric, 115
  stomal, 115
Ultracentrifugation, 85, 202
Uræmia, 13
  salt and water balance, 60
Urea, clearance, 10
  clearance test, 216
  concentration test, 216
  excretion of, 13
  in blood in diabetes, 140
Ureterocolostomy, 26
Uric acid, plasma, 204
  serum, 204
Uridine diphosphogalactose, 166, 167

Uridine diphosphoglucuronyl transferase, 169
Uridine diphosphoglucose (UDPG), 64, 78, 168
Urobilin, 64
Urobilinogen, 64, 68, 69
  Ehrlich's test for, 211
  fæcal, 73, 75
Uroporphyrin, 168, 192

Vagotomy, 114
Valine, 166
Van den Bergh reaction, 67
Vanillyl-mandelic acid (VMA), 154
Virilism, 150, 153, 221
Vitamins, 72, 182, 189–191, 201, 202
  deficiencies of, 120, 201, 202
  in the plasma, 190
Vitamin D (*see* Calciferol)
Vomiting, 14
  of pregnancy, 25
Von Giercke's disease, 167

Water, deficiency of, 50, 51, 54, 55, 58
  elimination test, 13
  inevitable loss of, 49
  intoxication, 56
  loss of, 50
  loss in surgical operations, 58
  metabolic, 49, 50
  of food, 49, 50
  treatment of deficiency of, 55
Water and salt deficiency, 58
  mixed, 58
Wernicke's encephalopathy, 181
Wilson's disease, 182

Xanthomatosis, 205
Xylose tolerance test, 122
1-Xylosuria, 166

Zollinger–Ellison syndrome, 116